ALSO BY JOHN WARDE

The New York Times Season-by-Season Guide to Home Maintenance

The Healthy Home Handbook

The Healthy Home Handbook

All You Need to Know to Rid Your Home of Health and Safety Hazards

John Warde

TIMES T BOOKS

RANDOM HOUSE

Copyright © 1997 by John Warde

Illustrations © 1997 by Edward Lipinski

All rights reserved under International and Pan-American Copyright Conventions.

Published in the United States by Times Books, a division of Random House, Inc., New York,

and simultaneously in Canada by Random House of Canada Limited, Toronto.

Library of Congress Cataloging-in-Publication Data

Warde, John.

The healthy home handbook :

all you need to know to rid your home of health and safety hazards /

John Warde. — 1st ed.

p.   cm.

Includes bibliographical references and index.

ISBN 0-8129-2151-8

1. Indoor air pollution—Management.

2. Home accidents—Prevention.

3. Dwellings—Remodeling.

4. House construction.

I. Title.

TD883.17.W37   1996

696 — dc20   95- 40980

Manufactured in the United States of America on acid-free paper

98765432

First Edition

Book Design by Robert C. Olsson

DISCLAIMER

In preparing this book, the author and editors have made every effort to achieve accuracy and to emphasize safe practices and techniques, but the reader assumes his or her own risk in performing all procedures and should approach the work with care. State and local building codes and practices may vary. Some of the construction practices described in this book may not be permitted by your local building department. The reader is responsible for complying with such codes and practices, including obtaining the required building permits. Construction materials, tools, skills, and site conditions also vary, and the reader assumes all risks relating to such matters. Always obey local and state laws, obtain required permits, follow manufacturers' instructions for operating tools, and observe safe work practices. No information mentioned in this book is intended to be construed as medical advice. If questions or problems arise, the reader should consult with a professional.

ACKNOWLEDGMENTS

This book was begun in 1992 and written over a period of four years. During that time many specialists and experts generously responded to my requests for information and advice, and without their help *The Healthy Home Handbook* could not have been completed.

In particular I wish to thank the following persons: Wally Anderson, Managing Director, Resilient Floor Covering Institute, Rockville, MD; Robert Axelrad, Director, Indoor Air Division, Office of Air and Radiation, U.S. Environmental Protection Agency, Washington, DC; William F. Behm, Managing Director, Childhood Lead Poisoning Program, Northeastern Pennsylvania Vector Control Association, Pittston, PA; Dr. Sue Binder, Lead Poisoning Prevention Branch, Center for Environmental Hazards and Health Effects, U.S. Department of Health and Public Services Centers for Disease Control, Atlanta, GA; Vince Collucio, National Lead Abatement Council, Olney, MD; Barry Dubin, Dumond Chemicals, Inc., New York, NY; Mark W. Earley, P.E., Chief Electrical Engineer, National Fire Protection Association, Quincy, MA; Anthony Giometti, Manager of Communications and Public Relations, American Society of Heating, Refrigerating, and Air-Conditioning Engineers, Atlanta, GA; Walt Gozdan, Technical Director, Rohm and Haas Paint Quality Institute, Spring House, PA; Roger M. Johnson, Vice President of Sales and Marketing, Nutech Energy Systems, London, Ontario, Canada; Sharon Kemp, President, DSK Environmental Safety Products, Beverly Hills, CA; Diane L. Maple, Director of Media Relations, American Lung Association, Washington, DC; Christopher Meehan, Buzzano Contracting and Radon Reduction Service, Newtown, CT; Bob Montessano, Sales

Acknowledgments

Manager, International Protective Coatings Corporation, Ocean Township, NJ; Ron Morony, Deputy Director, Office of Lead Abatement, Department of Housing and Urban Development, Washington, DC; David Murane, Radon Action Program, Office of Radiation Programs, U.S. Environmental Protection Agency, Washington, DC; Gary Nelson, Minneapolis Blower Doors, Minneapolis, MN; Philip M. Pecevich, Air Quality Research, Research Triangle Park, NC; Dr. Edward Vitz, Kutztown University, Kutztown, PA; Regis Wyse, Customer Information Representative, GMD Systems, Inc., Pittsburgh, PA.

At Times Books, my thanks go to Ruth Fecych, editor, and Naomi Osnos, design director, and their capable staffs. I am especially indebted to Edward Lipinski, illustrator and longtime friend. Most truly, I thank my wife, Kathy, for her undying faith, support, and perseverance.

CONTENTS

Contents

Appendix

INTRODUCTION

For any concerned homeowner the possibility that the air inside a house, the composition of building materials and products, and even the ground on which a house stands might contain harmful elements is a frightening prospect, and one of increasing concern.

Nor is the matter any longer news. Since the 1970s, beginning with the passage of the federal Clean Air Act and the push toward energy-efficient living fostered by the Arab oil embargo, evidence has piled up revealing the dangerous levels of toxic substances in the homes of most Americans—and the health risks implied. Along the way, other hazards have come to light, notably the prevalence of lead in older homes and the threat of fire and injury caused by inadequate electrical wiring.

But what to do about such problems? And perhaps more to the point in this era of media hype and overload, how bad are they really?

The Healthy Home Handbook is written to address these issues. Its aim is to provide straightforward, practical instructions—not merely what to do, but how to do it—for eliminating household pollutants, poisons, and safety hazards, and at the same time to present the most accurate and thorough information currently available on these subjects—sensibly, without bias, and without losing perspective in the face of zeal.

Part I, "Controlling Indoor Pollutants," contains individual chapters covering all of the harmful substances that can invade a house: asbestos, lead, radon, combustion products, biological contaminants, volatile organic compounds (VOCs), and noise (while not disease-producing, noise can be a powerful disruptive agent, as intolerable to healthful living as any conventional pollutant). There is even a chapter on improving ventilation. Each chapter answers in detail questions such as: What is the substance? How bad is the problem it creates? How can a home be tested? By what methods can the substance or its dangers be reduced or eliminated?

Hard facts supporting all sides of pollution issues are presented in an honest effort to display their full dimensions to thoughtful homeowners who are rightfully wary of half-truths. *The Healthy Home Handbook* does not underplay complicated scientific controversies, some of which surround even widely publicized air quality hazards—for example, asbestos and radon (neither of these substances has actually been proven responsible for a single human death, yet both are known causes of cancer). At the same time, potential dangers of pollutants are never minimized.

Regarding testing, hands-on methods for amateurs are detailed thoroughly, together with guidelines for locating and hiring qualified professionals and what to expect from their services. Testing for indoor pollutants frequently is difficult for amateurs and professionals alike, owing to the small quantities of substances involved and the fact that dosage limits for many compounds are unknown. *The Healthy Home Handbook* makes clear the practicalities of testing and how to gauge the accuracy and importance of results.

Of course, techniques for mitigating—reducing or eliminating—harmful substances are the core of each chapter. Clear instructions for performing a complete range of fixes is provided for each pollutant, from simple stopgap measures (where they will work) to full-scale renovations where appropriate. Throughout, the focus is on practicality, effectiveness, and safety. Methods are those used by professionals and recommended by such authorities as the U.S. Environmental Protection Agency (EPA), the federal Occupational Safety and Health Administration (OSHA), the Centers for Disease Control (CDC), the American Lung Association (ALA), and the National Association of Home Builders (NAHB), to name a few.

Every effort is made to explain procedures in full, so that homeowners of all skill levels can benefit and so that methods of mitigating hazards can be justly compared. No two homes are the same; the thorough explanations and the breadth of information in *The Healthy Home Handbook* are intended to enable readers to gain expert knowledge in lieu of experience, so they can make their own decisions, taking into account individual circumstances and conditions.

Part II, "Minimizing Safety Hazards," covers causes of physical injury in houses. Less controversy surrounds these issues, therefore information presented in this section focuses on identifying hazards comprehensively and providing clear remedies. Protecting against fire and falls, the two most common causes of household accidents, receives specific attention. For example, besides instructions for carrying out thorough room-by-room inspections to search for household fire hazards, complete comparisons, buying advice, and maintenance tips are presented for fire extinguishers and smoke detectors, along with details of creating a family escape plan in case of actual fire.

Household electrical systems also are fully described and explained, with emphasis both on finding and correcting faults and on upgrading old wiring. Instructions are provided for mapping electrical circuits to avoid dangerous overloads, detecting hidden electrical leaks, and estimating electrical power needs for an entire house to determine whether additional circuitry is needed.

Other chapters in Part II address childproofing and providing comfortable accommodations for the elderly and disabled. The childproofing chapter details ways to protect toddlers from injuries—especially bumps, scrapes, and falls—and ways of preventing access to hazards such as electrical sockets and household poisons. "Creating a Barrier-Free Home" explores practical methods of eliminating common barriers to accessibility in homes; for example, installing an entrance ramp, replacing round doorknobs with levers that are easier to operate, and customizing kitchens, bathrooms, bedrooms, and living areas to make them easier to navigate and occupy.

At the end of the book there is an appendix that contains the latest findings on three controversial topics: electromagnetic fields, seasonal affective disorder (wintertime depression caused by lack of daylight), and multiple chemical sensitivity, variously known as total allergy syndrome, environmental illness, and twentieth-century disease. The final section, "Where to Find Help," lists additional sources of information, organizations, and products.

Much research, deliberation, and verification has gone into *The Healthy Home Handbook*. Although no book, especially one about

pollution, can be complete—or perhaps even current—for long, it is
hoped that the information this work contains will prove valuable
to many homeowners and their families and serve a useful purpose
as a lasting guide and reference for improving household health and
comfort.

ILLUSTRATIONS

Asbestos

Lead

Radon

Combustion Products

Biological Contaminants

Volatile Organic Compounds

Improving Ventilation

Controlling Noise

Protecting Your Home from Fire

Upgrading Old Wiring

Illustrations

Avoiding Falls and Other Injuries

Childproofing

Creating a Barrier-Free Home

PART I

Controlling Indoor Pollutants

Asbestos

What Is It?

ASBESTOS, CHIEFLY NOTED FOR ITS INSULATIVE QUALITIES—especially its resistance to fire and heat—nowadays is familiar only to a dwindling generation of adults. It is a fibrous material, usually white or gray in color, rather heavy when woven into cloth (asbestos pot holders and oven mitts were popular in the 1950s and 1960s), but light and fluffy when used as furnace or ceiling insulation. The cottony snowflakes falling on the set of the movie *Holiday Inn* as Bing Crosby crooned the legendary "White Christmas" reportedly were pure asbestos. Crosby died in 1977 of lung cancer.

Technically, asbestos refers to the crystalline form of several iron- and magnesium-enriched silicate minerals. Six are considered major forms: chrysotile, crocidolite, amosite, anthophyllite, actinolite, and tremolite. Approximately 95 percent of the asbestos used in building materials and other commercial products is chrysotile; crocidolite and amosite account for nearly all the rest.

Chrysotile is sometimes called white asbestos. It belongs to the family of minerals called serpentine and is characterized by fibers that are curly, silky smooth, and flexible. A pound of chrysotile can yield thirty thousand feet—about six miles—of fiber; individual strands are about a thousand times smaller in diameter than human hair and have the tensile strength of piano wire.

The other asbestos varieties belong to the mineral family called amphibole. Their fibers are straight, rigid, and sharp like needles. Crocidolite, sometimes called blue asbestos, has long, coarse fibers that are especially resistant to acids. Amosite, or brown asbestos, features even longer coarse fibers having high iron content, hence

3

their color. Amosite is seldom spun; and while all asbestos insulates against electricity, the iron in amosite renders it less effective as an electrical insulator than varieties having less or none of the metal. Generally, amosite is used as a binder in insulation products.

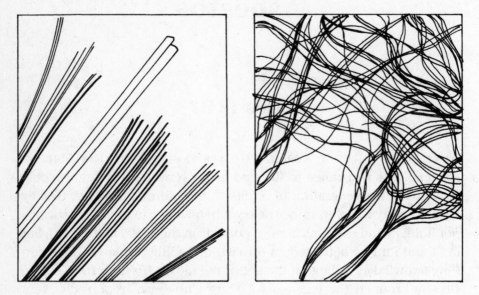

Amphibole fibers   *Chrysotile fibers*

Illustration 1. Microscopic views of asbestos types

Both anthophyllite and actinolite feature short, brittle fibers unsuitable for spinning. They, too, generally are ground up and mixed into insulation products and plastics. Tremolite has high resistance to chemicals and is an excellent material for filters.

Asbestos has been used since prehistoric times. Stone Age potters added asbestos to clay, probably as reinforcement against breakage. The Egyptians wove it into funeral shrouds that would not decompose; the Romans wove it into tablecloths and napkins that could be thrown into a fire for cleaning. Centuries later, Marco Polo reported seeing cloth in Genghis Khan's northern provinces that probably was made of asbestos.

The largest modern asbestos deposits are in Canada, southern Africa, and China. The discovery of the Canadian deposit (located in Quebec) in 1878, coupled with advances in the Industrial Revolution, brought asbestos to prominence. By the 1970s, when the mineral's use in America became regulated by the federal government through the Occupational Safety and Health Administration (OSHA) and the Environmental Protection Agency (EPA), the material could be found in more than three thousand manufactured products.

Disease associated with exposure to asbestos also has been recognized since the first century A.D., when Pliny the Elder noted ill health and early death among Roman slaves who wove asbestos fibers. But it was not until the 1920s that asbestos was named the culprit in such maladies, and only in the 1930s was asbestos recognized as a carcinogen.

The noncancerous maladies attributed to asbestos comprise a disease called asbestosis. Symptoms include coughing, excessive production of sputum or phlegm, frequent lung infections, and a dry, crackling sound in the lungs during breathing. Increasing shortness of breath, weight loss, and heart failure are also signs of the disease. The cause of asbestosis is scarring of the lungs after prolonged inhalation of asbestos fibers. The scarring actually is the encapsulation of fibers by small, benign fibroid tumors that eventually grow together. Victims of asbestosis usually die within fifteen years of the onset of symptoms.

The first cancer variety recognized in asbestos workers as being caused by asbestos was lung cancer. Among persons heavily exposed, the mortality rate from lung cancer is about 25 percent, according to some studies; among the general population, the mortality rate from lung cancer is only about 5 percent. An asbestos worker who smokes, or who did so during the time of exposure, is about ninety times more likely to develop lung cancer than a smoker who has never been exposed to asbestos.

During the 1960s another, rarer cancer was also noticed. This is mesothelioma, a cancer of the chest and abdominal cavity linings, or pleura. Symptoms include shortness of breath and pain in the chest wall or abdomen. Eventually, tumorous growth can occur that may crush the lungs or abdominal organs, or protrude outside the body. Seven to 10 percent of deaths among asbestos workers are attributed to mesothelioma. Most cases develop in workers exposed to crocidolite fibers.

Mesothelioma became a household word when actor Steve McQueen died of the disease in 1980. Exposure to asbestos is practically the only known cause of this kind of cancer; McQueen's exposure may have occurred during the 1950s, when he was a merchant sea-man—ships are notorious for containing huge amounts of asbestos insulation—or later, when he was an avid amateur race car driver and mechanic (asbestos fibers are a chief component of automotive brake linings even today, and can easily be inhaled during servicing). Unnervingly, mesothelioma has been discovered in people with little direct exposure to asbestos—wives and children of asbestos workers and people living near asbestos factories.

Symptoms of all asbestos-related disease generally take ten to forty years to appear after the first exposure. This helps explain why it has taken so many years for health researchers to recognize asbestos as a significant hazard. Unfortunately, it also suggests that by the time symptoms are noticed in a victim the injury from asbestos has long since occurred and the opportunity to halt exposure or heal the damage has long passed.

How Bad Is the Problem?

Until the early 1960s and the rise in cases of mesothelioma, asbestos-related disease was judged to be primarily, if not exclusively, an occupational hazard; that is, a problem among factory workers exposed to vast amounts of asbestos fibers for long periods each day and for many consecutive years. Particularly vulnerable were shipyard workers, pipe fitters, and (as in ancient times) employees of textile mills that manufactured asbestos cloth.

But during the 1960s and 1970s, as research focused on the effects of asbestos among the general public, it seemed apparent that asbestos also posed a threat to ordinary people. Homes and offices, many of which are still in use, contained millions of tons of asbestos pipe insulation, floor and ceiling tiles, wallboard and plaster materials, fireproofing, electrical insulation, and roofing.

Friable asbestos products—any items or materials that can release asbestos fibers into the air—were and are of principal concern. Research data so far shows that asbestos is a health hazard only when inhaled. There is no known safe level of exposure—the current permissible exposure limit, or PEL, set by OSHA in 1986 is 0.2 fibers of asbestos per cubic centimeter of air inhaled during an eight-hour period—but to date no research indicates risk from asbestos in products where fibers are entrained; that is, entrapped or embedded so they cannot escape.

United States federal and state regulations concerning asbestos control regard all friable asbestos as equally hazardous. The Clean Air Act, passed by Congress in 1970, was the first such legislation. The bill gave the EPA authority to regulate hazardous air pollutants, and asbestos was one of the first three substances the EPA intended to regulate (the other two were radon and lead; both are discussed in other chapters of this book).

This was followed in 1973 by an EPA ban on the spraying of asbestos materials and by the enactment of the National Emission Standards for Hazardous Air Pollutants (NESHAP), which requires removal from public buildings any asbestos that would be disturbed or damaged during future renovation or demolition. In 1989, the EPA announced a three-stage phasedown program designed virtually to end the use of asbestos. This was never fully implemented owing to the number of manufacturers who voluntarily stopped producing materials and goods containing asbestos. Today, the only federal ban on asbestos in manufacturing bars the use of asbestos in newly developed (that is, never-before-invented) products.

However, since the late 1980s there has been heated controversy about whether or not chrysotile asbestos, the variety most commonly used and the one whose fibers are distinctly softer than the others, may be less harmful than amphibole asbestos, especially to the general public. Support for this argument stems chiefly from statistical evaluations performed by Dr. Brooke T. Mossman, a research pathologist at the University of Vermont, and a group of colleagues. Their work has focused on twenty-year studies beginning in the 1960s revealing that while rates for all asbestos-related disease rose steadily during the period, the new victims mostly were older men who had been heavily exposed to asbestos as workers years before. By contrast, rates of asbestos-related disease among women, whose exposure typically was from environmental sources, not occupational ones, remained stationary and cases were all but confined to mesothelioma.

Why didn't the increase in disease appear in the women as it did in the men, and why was mesothelioma prevalent while other asbestos-related diseases were virtually excluded? Mossman's conclusion, published in 1989, was that it was because the women studied were exposed primarily to chrysotile asbestos, practically the only kind found in large amounts outside of industrial settings. That the figures for women also remained stationary despite an increase of chrysotile asbestos in the environment (due to growing use of asbestos in consumer products until it became regulated) further suggests the material's relative harmlessness. According to Mossman, the reason mesothelioma was virtually the only asbestos-related disease found was because conditions that typically produce the other diseases—high, prolonged concentrations and contact with amphibole asbestos varieties—simply were not present.

Mossman's studies are corroborated by European medical findings showing that chrysotile's flexible, smooth fibers are expelled by the lungs before they can cause serious harm, while the spiny fibers of the amphiboles lodge permanently. Only the United States and South Africa, among the world's industrial nations, continue to lump chrysotile asbestos with the amphiboles as potent causes of cancer at nonoccupational levels. Still, with the controversy formally unresolved and with no known safe level of asbestos exposure, homeowners with the opportunity to rid their houses of an asbestos threat, regardless of degree, may be well-advised to do so, but with the following caveat: According to an article published in *Consumer Reports* (July 1995), there has never been a single case ever reported of an asbestos-related disease attributable solely to exposure to home building materials; and even the risk of cancer to average citizens, extrapolated from the known disease rates of professional asbestos workers, is minuscule—about 1 chance in 100,000 over a lifetime period of exposure.

Testing

If your home or apartment was built or remodeled between 1920 and 1978 it probably contains asbestos in some form. Places where asbestos is most likely to be found are:

- heating system ducts
- door gaskets on furnaces and wood- or coal-burning stoves
- hot water and boiler pipe insulation

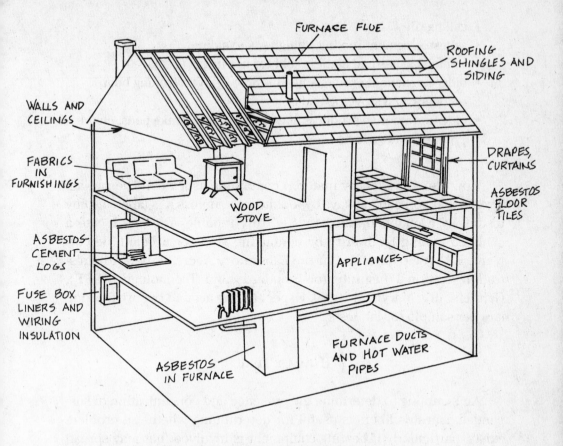

Illustration 2. Sources of household asbestos

- fuse box liners
- lamp socket collars
- insulation on electrical wiring
- paper lining around built-in stoves and dishwashers
- small appliances made before 1979, like freezers, ovens, toasters and hair dryers
- soundproofing or decorative material sprayed on walls and ceilings
- resilient flooring
- cement sheet, millboard, and paper used as insulation around furnaces and wood-burning stoves
- patching and joint compounds for wallboard
- textured paints

- ceiling tiles
- asbestos cement roofing, shingles, and siding
- artificial ashes and embers sold for use in gas fireplaces
- older household products like fireproof gloves, ironing board covers, and stovetop pads
- older automotive products like brake linings, brake pads, clutch plates, and gaskets

How can you identify material containing asbestos? The sad answer is that you can't always by looking at it, unless it is labeled. Only testing by a qualified laboratory is truly reliable. You can locate a qualified testing laboratory by contacting your local health department, EPA regional office, or the Laboratory Accreditation Administration, National Institute for Standards and Technology (NIST), Gaithersburg, Maryland 20899. Expect to pay around twenty-five dollars per sample for lab tests.

How to Test

Air sampling to determine the presence and concentration of suspended asbestos fibers is useful for determining whether a problem exists, particularly if heating equipment is involved that may spread asbestos throughout a house via ductwork. However, such testing can only be done by a professional using special equipment—generally either a sampling pump or a computerized digital fiber monitor—and is seldom done in private homes unless asbestos is first identified visually. Air sampling should be done prior to any attempt at correcting an asbestos condition and must be done afterward to comply with EPA and other recommendations.

Taking samples of suspected asbestos-containing materials to send to a laboratory also should be done by a professional, but a careful homeowner can do it by learning the correct techniques and by exercising extreme caution. If done incorrectly, sampling can be more hazardous than simply leaving alone material that is otherwise intact. A professional knows what to look for and can take samples safely, without creating a risk where none exists.

To have your home inspected and sampled professionally, call a testing service or an asbestos abatement contractor. Both are listed in the yellow pages; look under "Environmental and Ecological Services" and, of course, "Asbestos." Some asbestos professionals offer inspections, testing, and removal or other corrective action. The firm you hire to inspect and/or test your home should not be the same as or affiliated with the company that undertakes corrective action; it is better to use different firms to avoid being defrauded.

To select a testing service or asbestos contractor, start by calling your local board of health or department of consumer affairs. These offices can tell you what licensing and certification requirements must be met in your area, and in some cases they may have lists of qualified companies you can contact. While the federal government certifies training courses for asbestos professionals (many of these courses are required for licensing or certification in various states), there is no federal certification program. Instead, licensing and certification of testing companies and asbestos contractors are handled at state and local levels.

Ask asbestos professionals about their training and, as when selecting any professional service, be sure the company or individual meets at least minimum criteria for qualification and can supply proof of Workman's Compensation insurance. Find out how long the business has been operating and get a list of prior clients. Be aware that licensing often merely involves paying a fee; the best contractors will have passed some kind of screening test.

Performing Tests Yourself

You can gather samples yourself of material that you think may contain asbestos, but you must learn to do so safely and be careful not to make mistakes. The important factors are taking care not to release asbestos fibers into the air and not to inhale them yourself. Do not sample material in good condition that is likely to remain undisturbed. Sample only broken material or material destined to be disturbed by remodeling or for some other reason. Following are some clues on what to look for; for help in identifying asbestos on plumbing and heating equipment, call for an examination by an experienced plumber or heating company technician.

Before taking a sample, contact the lab to which it will be sent and obtain any special instructions. Unless other techniques are ad-

A GUIDE TO IDENTIFYING HOUSEHOLD ASBESTOS

• Asbestos surrounding metal heating ducts is usually an off-white color and may resemble corrugated cardboard. Some ducts are made entirely of asbestos.

• Asbestos surrounding plumbing—usually pipes connected to steam-heating radiators—may be covered with canvas. Where any asbestos is visible it is usually white and crumbly in texture.

• Asbestos gaskets on wood stoves and furnaces, and protective wall panels near stoves, usually are grayish white in color and resemble stone. Close inspection reveals their fibrous qualities.

• Asbestos on walls and ceilings gives these surfaces a textured, even lumpy appearance. Old acoustical ceilings whose texture resembles cottage cheese often have been sprayed with asbestos.

• Asbestos used as insulation in attics or walls usually is white or off-white and resembles packed cotton.

• Asbestos used in plaster, joint compound, resilient flooring, and roofing and siding materials generally cannot be seen. You should suspect the presence of asbestos if they were manufactured before 1980; materials manufactured more recently can be considered safe.

vised, first shut off any heating or cooling system in the room that might circulate released fibers in the air or through ducts. Put on disposable gloves. Wear a dust mask having a high-efficiency particulate-arresting air filter (called a HEPA filter) labeled effective against asbestos particles by a major testing organization like OSHA or NIOSH (the National Institute of Safety and Health). These masks are available for around twenty dollars from suppliers of safety equipment and materials (check the yellow pages under "Safety Clothing and Equipment"). A mask without a HEPA filter offers no protection against asbestos.

Work alone in the room. Place a plastic sheet on the floor beneath the material to be sampled to contain any that falls. Wet the material with water mixed with a small amount of liquid dish detergent (about a teaspoon of detergent per quart of water) to reduce the release of asbestos fibers; apply it with a spray bottle. Cut or pull off a small piece of the material that includes its full thickness or depth (for example, when sampling duct wrapping cut all the way through the material until you reach the duct). Use a sharp penknife, a vegetable peeler, or a similar tool. Place the sample in a clean container like a jar, pill vial, or 35-mm film canister and fasten the lid tightly.

Unless you are simply pulling material from a crumbling area, cover the hole made by sampling with a piece of duct tape to prevent any fibers from escaping. Afterward, wipe the area with a damp paper towel that you then deposit on the plastic sheet.

Carefully fold the corners of the sheet toward the center and roll the sheet up for disposal. Wipe the floor and any areas that may have received fibers, including the outside of the sample container, with additional damp towels. Finally, place the waste and your disposable gloves in a plastic bag, seal it, and dispose of it in the trash. State and local procedures are quite strict about disposing large amounts of asbestos, but the amount contained in household sampling waste is likely to be too small for special action. If in doubt, call your local health department or sanitation bureau for advice. Wrap the sample securely and mail it to the lab according to its instructions.

Getting Rid of Asbestos

Deciding on a Course of Action

If asbestos in a house is in good condition—the fibers are completely contained—and will not be disturbed, the best course of action is simply to do nothing; leave the material alone. Disturbing asbestos by attempting removal risks creating a problem where none exists and in addition constitutes unnecessary expense.

Even if asbestos material is slightly damaged, if it is located in an isolated area of the house where people are unlikely to go—for example, a crawl space or an unfinished basement corner—the best strategy for coping is simply to limit access to the area, perhaps post a warning sign nearby, and leave the asbestos undisturbed.

On the other hand, if asbestos is or has become friable, with fibers loose and able to float about in the air, or if asbestos in good condition is located where it can be easily disturbed or regularly encountered, steps should be taken to eliminate the hazardous condition. There are three ways to do this: removal, enclosure, and encapsulation.

Removal

Removal is the most extreme answer to an asbestos problem. If there are alternatives, pursue them first; they are likely to be easier to accomplish and less expensive. However, removal does have certain advantages: It is a permanent solution to an asbestos condition; it ends the problem forever; and there is no need for maintenance or for monitoring changes in the condition of the material or of any protective barriers surrounding it. Removal also enables renovation or remodeling to be carried out safely at any time.

Disadvantages, besides difficulty and expense, are that new material often must be installed to replace asbestos that has been removed, and, more important, that removal, unless performed expertly, can create a far worse condition than it is intended to cure. A bit of asbestos wrapping that enters a heating duct, for instance, can contaminate an entire home in minutes. So can vacuuming dust from a work site with an ordinary household vacuum cleaner or failing to adequately seal off a site while removal is taking place.

Once removed, the asbestos can present another problem: disposal. Even in relatively small amounts, asbestos must be handled as hazardous waste to comply with local restrictions. Not only does this mean special handling to bundle the material properly but it means paying expenses (usually high) for special hauling and dumping in a designated landfill.

Finally, of course, there is the potential health danger to workers actually performing removal and to occupants of the house during the project, though they should not be at home. Few amateurs have the skills to work safely with asbestos or can justify the cost of obtaining the protective clothing and equipment needed to insure exposure-free handling of the material. In the end, where removal of asbestos is necessary, the job is best handled by qualified, experienced professionals. And in a growing number of areas, hiring licensed or certified asbestos-removal technicians is required by law.

Enclosure

Surrounding asbestos with other material or with a partition like an airtight wall or ceiling to contain fibers usually is less expensive than removal (at least in the short run) and sometimes can be done by a careful amateur. For example, heating ducts or pipes covered with asbestos can be wrapped with special tape, described below, or boxed within a framework of lumber and wallboard. Before boxing ducts, make sure their subsequent inaccessibility does not violate local building and fire-safety codes.

Wrapping tape for ducts and pipes is usually called lagging cloth and, like HEPA filters, can be obtained from safety equipment and materials suppliers. Lagging cloth usually contains fiberglass and is applied after first dipping it in water to activate the adhesive and then squeezing out the excess moisture.

To wrap a pipe or duct, begin by examining the damaged area to determine the length of lagging cloth needed. (Remember, covering undamaged asbestos in an out-of-the-way location is not necessary.) Wear the types of clothing and take the precautions described earlier for testing. Be especially careful when approaching damaged asbestos that you do not step on or brush against pieces or dust containing fibers that may have fallen to the floor or onto nearby horizontal surfaces. Removing such material may require professional handling and is described further on.

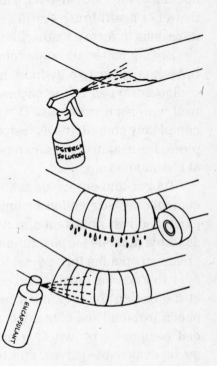

Illustration 3.
Applying lagging cloth

Spray the damaged asbestos with a mixture of water and dish detergent (as described for testing) to reduce the ability of any fibers to become airborne. After cutting the cloth to length, wet it thoroughly, squeeze it only enough to keep it from dripping, and then wrap it around the damaged area like a bandage. On hot pipes, most tape dries in two to four hours. On cold pipes, drying time usually is between eight and twelve hours.

When the tape is dry, spray or brush it with a penetrating asbestos encapsulant (a sealer available where lagging cloth is sold) until the fabric is saturated. When it dries, apply a second coat. If a tough, damage-resistant surface is needed, apply an asbestos encasement coating to the cloth instead. This is a mastic that can be spread with a trowel or a stiff brush (professionals usually use an airless sprayer); it hardens to form a leatherlike coating.

Boxing materials containing asbestos can be done using one-by-three lumber or two-by-fours as studs and then covering these with wallboard, taped at the seams as you would do when building any wall with such materials. The difference is that the enclosure must completely contain the asbestos material. Boxing for vertical heating pipes, for instance, must extend from floor to ceiling and be caulked at top and bottom.

Tongue-and-groove or shiplap paneling can also be used. Not acceptable are suspended ceilings with tiles or panels that slide into place and rest unfastened in a frame. Lights in any ceiling covering asbestos must be surface mounted, not recessed, as the latter would create an opening through which fibers could escape.

After making repairs, and especially if debris and dust are present at the outset, the area must be thoroughly cleaned. Where cleanup needs are small and contained, you can perform the task without special equipment by working carefully as follows: Wearing a HEPA mask, disposable gloves, and other clothing that covers as much of your skin as possible, spray the surfaces to be cleaned with water mixed with dish detergent. Then mop or wipe them with a rag, a plain sponge, or a sponge mop.

Vigorously rinse the items under running water after the first pass, then repeat the procedure one or two additional times. When finished, dispose of the items used for cleaning if possible; otherwise, vigorously rinse them as before. Try to change clothes, including shoes, before leaving the area so you do not distribute fibers that may be clinging to them. Wash work clothes promptly, separated from other clothing.

The cleanup appliance of choice is a HEPA vacuum. This is an industrial vacuum cleaner designed for removing hazardous particles

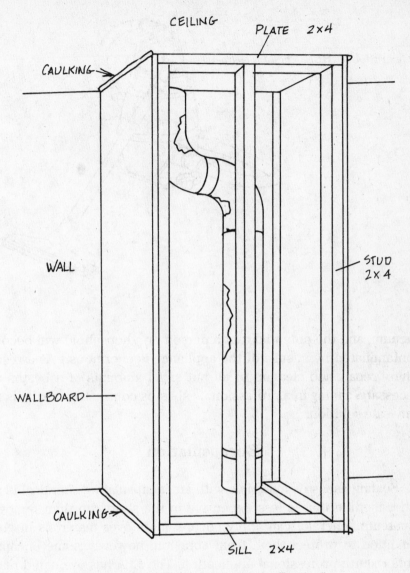

CEILING

PLATE 2×4

CAULKING→

WALL

STUD 2×4

WALLBOARD—

CAULKING→

SILL 2×4

Illustration 4. Enclosing a pipe by "boxing"

and liquids. It features, among other things, a HEPA filter capable of containing asbestos fibers within the machine so they are not released into the air. Unfortunately, amateurs stand little chance of obtaining a HEPA vacuum. The least expensive models cost around $500 and few, if any, companies rent them. (The reasons are obvious: Training is required to properly operate, empty, and clean a HEPA

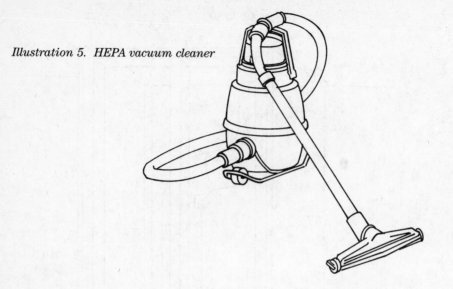

Illustration 5. HEPA vacuum cleaner

vacuum, and the potential that a person or a household will become contaminated by misuse of the appliance is enormous.) As a result, where repair and cleanup of all but small amounts of asbestos are necessary, calling in a professional asbestos contractor usually is the only safe solution.

Encapsulation

Sealing asbestos materials with an encapsulating chemical is another method that is less expensive in the short run than removal. Amateurs generally can apply encapsulants over materials in good condition to protect them from abrasion; however, some encapsulants require professional application. The EPA has evaluated many encapsulants; for a list, telephone the EPA's Office of Toxic Substances, TSCA Assistance Office, at (202) 554-1404.

Some high-quality latex paints can be used to encapsulate granular, cementlike asbestos materials, according to the EPA. Such paint must have at least 60 percent vehicle (solvent) content by weight and at least 25 percent resin. Elastomeric wall coatings designed as vapor barriers are also effective. Spraying is the best method of application because it covers large areas quickly and because there is less of a tendency for the process to rub off particles containing fibers. But a roller or even a paintbrush also can be used.

To encapsulate a surface with latex paint or an elastomeric coating using a brush or roller, apply it in two thick layers with time allowed for drying in between coats. If spraying, use airless equipment. Apply a light coat first, then a heavier, full coat. Some experts recommend applying the second coat at a ninety-degree angle to the first to insure even coverage.

Provided an asbestos-coated wall or ceiling is in good condition, another viable encapsulating material is flexible veneer wallcovering. This product, intended for restoring plaster surfaces, resembles thick wallpaper and is available in varieties made of spun fiberglass, woven fiberglass, and plaster-saturated burlap.

Spun-fiberglass wallcovering has no distinct texture and is a good choice for encapsulating surfaces to be painted or covered with ordinary wallpaper. To apply it, the wall or ceiling is first coated with paint recommended by the manufacturer. Then, while the paint is still wet, the material is pressed against it and covered with a second layer. When dry, the wall is ready for ordinary painting or wallpapering.

With the other materials, special adhesive is applied to the wall or ceiling instead of paint. The material then is pressed against it and flattened with a roller. Because the material is decoratively textured and colored, no further treatment is necessary.

Enclosure and encapsulation have another advantage besides avoiding the problems of removing asbestos: No new material need be added to substitute for asbestos that has been removed. But both remedies require continual monitoring after the job is completed, and eventually removing the asbestos may be necessary anyway, wiping out any savings gained. For your own reference and for future occupants of your house, keep a record of areas where asbestos has been enclosed or encapsulated.

What to Do with Nonfriable Asbestos Materials

Asbestos whose fibers are entrained—that is, are embedded in another material so they cannot escape into the air—presents little, if any, problem. Common building materials containing entrained asbestos are roof and siding shingles, and resilient flooring, which today is the most prevalent source of household asbestos. Sanding, scraping, cutting, or breaking these materials can release fibers. As a result, many communities permit only qualified asbestos contractors to remove them and recommend leaving them in place and simply covering them with new nonasbestos material whenever possible.

Resilient Flooring

With care, amateurs can remove resilient flooring. Any that was manufactured before 1986 is likely to contain asbestos; if you can be sure that the flooring you plan to remove was manufactured later than that date there is no problem unless earlier flooring exists underneath it.

The first step, of course, is to determine that removing the asbestos-containing flooring is necessary. Generally, the only situations where removal is required are when flooring has curled, is extensively broken, or is poorly bonded. Most of the time, old resilient flooring in reasonably good condition can be prepared by filling in low spots (including embossed patterns) with latex floor-leveling compound, available from flooring supply stores and many home centers. When the compound is dry, new flooring can be installed directly on top of the old.

Another solution is to install new underlayment, usually quarter-inch-thick lauan panels, over the existing flooring to create a smooth base. This can be done only if the existing flooring rests on a wood subfloor (a concrete floor prevents nailing the underlayment to fasten it), but it is the best way to level an uneven floor. Leveling compound can be applied to the existing flooring first; never sand or scrape flooring to reduce high spots.

Sheet Vinyl

Removing sheet vinyl flooring requires two people. One person must spray the area being removed continuously with water mixed with detergent while the other performs the removal procedures. The Resilient Floor Covering Institute, a trade association of vinyl flooring manufacturers, recommends using sixteen ounces of liquid dish-

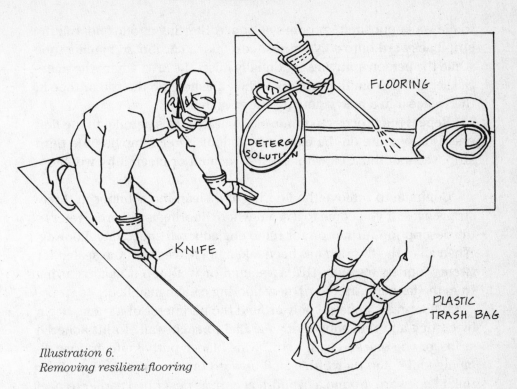

FLOORING

DETERGENT SOLUTION

KNIFE

PLASTIC TRASH BAG

Illustration 6.
Removing resilient flooring

washing detergent with one gallon of warm water and applying it with a garden sprayer. This minimizes airborne asbestos particles.

All of the measures described earlier for testing and enclosing asbestos materials should be followed. While vacuuming before and after removal with a HEPA vacuum is strongly advised, working carefully to avoid creating airborne particles and then cleaning the area by wet-mopping during the operation, upon completion, and again twenty-four hours later, when airborne particles have settled, is an effective substitute. As mentioned earlier, never substitute an ordinary vacuum cleaner for a HEPA vacuum: Asbestos picked up by the vacuum will pass through the liner and be blown into the air by the machine's exhaust.

Following are removal procedures recommended by the Resilient Floor Covering Institute:

Start removal by prying away all of the moldings and binding strips around the floor's perimeter. Next, if the flooring is loose-laid—having no adhesive underneath—use a linoleum knife or utility knife to cut a full-length strip of the material about six inches wide parallel to the wall opposite the room's entrance. Starting at one end, roll the strip backward onto a cylindrical core like a cardboard mailing tube while the person spraying continually mists the area where the material separates from the subfloor. When finished, tie the roll or tape it; then place it in a heavy-duty plastic trash bag.

Repeat the process to remove two more strips parallel to the first one. Do not walk on the exposed subfloor. After removing the third strip, vacuum the area with a HEPA vacuum or clean it by wet-mopping.

Continue to remove the flooring and clean the subfloor in three-strip stages. If you come to a seam where flooring is adhered, refer to the description further on for removing adhered sheet vinyl flooring. When all of the flooring has been taken up, leave the room sealed for twenty-four hours, then HEPA vacuum or wet-mop it again starting from the far end. Afterward, new flooring can be installed.

If flooring is adhered only around the perimeter of a room, begin by cutting and removing strips parallel to each wall. To lift adhered strips, pry up one end with a stiff, wide-bladed putty knife or drywall-taping knife, and then either pull upward on the strip at an angle or roll it backward around a cylindrical core as described earlier to peel the flooring away from the adhesive underneath.

If any of the flooring's foam backing sticks to the adhesive, try pulling the strips loose from the opposite end. If this doesn't work, scrape the foam from the adhesive with the putty or drywall-taping knife (be sure it is wet throughout), and use the tool to help separate the flooring from the adhesive as you continue to pull or roll it.

Roll and tie each strip and deposit in a heavy-duty trash bag as described. When all of the adhered flooring has been removed, proceed as for removing loose-laid flooring.

If vinyl sheet flooring is fully adhered, remove strips using the lifting technique for perimeter-adhered flooring, but start at the rear of the room as you would to remove loose-laid flooring.

Sometimes, particularly when several layers of flooring are present or if the floor has a thick foam backing requiring much scraping, a more practical method of removal is to pry up the underlayment to which the resilient flooring is adhered. Whether or not this is feasible depends on whether any underlayment exists beneath the flooring (on some floors it does not), and whether you can find the seams where underlayment panels meet.

Removing a full-depth strip of flooring from around the perimeter of the room should answer the first question and reveal at least one underlayment seam. Starting at a corner, lift the underlayment a short distance with a pry bar to get a sense of that panel's shape. Then mark where you think its seams are and remove strips of flooring covering them. When the seams are exposed, pry up the panel completely, trying not to break it. (If it does break, slice through the flooring at the break, spray the area with detergent solution until it is thoroughly wet, and then proceed.)

Lean the panel against a wall and HEPA vacuum or mop the area from which it was removed. Finding the seams of adjacent panels should be easier than locating the seams of the first panel; continue removing panels until all the flooring has been lifted.

Finish the job by HEPA vacuuming or wet-mopping twenty-four hours later. Wrap the underlayment panels in heavy plastic sheeting (six-mil thickness is recommended), label it, and dispose of it according to instructions from local building or sanitation authorities. Naturally, before installing new flooring, you will have to install new underlayment.

Resilient Tile

Removing resilient tiles requires heating them to soften the adhesive underneath. Flooring professionals often use special radiant-heating devices; amateurs can substitute a heat gun or heavy-duty hair dryer that blows hot air.

Prepare for removing tiles as was described for removing sheet vinyl flooring. Clear the room, isolate it from other areas of the house, and remove any baseboard moldings or binding strips. Least-trafficked areas of flooring—for example, out-of-the-way corners—generally are easiest to lift because they have not been pressed against the adhesive as often or as hard, so start by removing these. The goal is to remove individual tiles without breaking them; this keeps the release of asbestos fibers to a minimum.

To remove a tile, heat the surface for about thirty seconds while prying beneath one edge with a wide-bladed putty knife or a drywall-taping knife. When the knife slides beneath the edge, twist the blade to pop the tile free from the softened adhesive. Place removed tiles in a heavy-duty trash bag for disposal.

The first tile is usually the most difficult to lift. With it out of the way, getting at the other tiles is easier and heating may not always be necessary. To remove difficult tiles, heat them thoroughly and try hammering the knife beneath them while holding the blade at an angle of about twenty-five degrees to the floor. Another solution is to build a heat box to focus the hot air from the heat gun or hair dryer. To build a heat box, cut five pieces of half- or quarter-inch-thick plywood and fasten them together with screws or nails to form a box whose perimeter is the same as or a little larger than that of a tile. Drill a hole in the bottom of the box to accommodate the nozzle of the heating appliance. To use the box, place it upside-down over a tile. Insert the heat gun or hair dryer and operate it for about thirty seconds. Then remove the box and quickly pry up the tile before the adhesive rehardens.

Remove tiles from a floor in such a way that you do not have to walk on bared areas of residual adhesive, which contains asbestos fibers. Residual adhesive must be smoothed if new vinyl tiles requiring asphaltic adhesive are to be installed in place of the removed tiles; if new tiles will require latex, resin, or epoxy adhesive the residual adhesive must be removed.

On concrete floors, smoothing residual adhesive is done by wet-scraping the surface as areas of about thirty square feet are exposed.

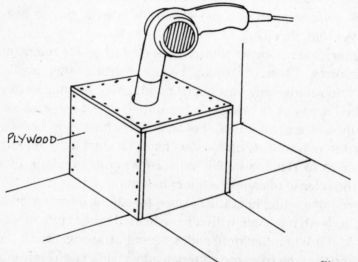

PLYWOOD

Illustration 7.
Homemade heat box for softening tile

While the job can be performed by one person, two can perform it better and more quickly.

You will need a garden sprayer filled with a detergent-and-water solution—one ounce of dish detergent to one gallon of warm water. While one person continually mists the area of adhesive to keep it wet, the other person scrapes away all ridges and high spots with a stiff-bladed putty knife or long-handled floor scraper to leave a thin, smooth film.

Place the fragments of adhesive in a heavy-duty trash bag and then carefully sweep and mop the floor while it is still wet. If possible, vacuum the floor with a HEPA wet-dry vacuum instead. After completing the floor, wait twenty-four hours, then mop or vacuum it again. Before installing new flooring, cover the residual adhesive film with trowelable underlayment compound or self-leveling cementitious underlayment and allow it to harden completely.

If residual adhesive must be removed, consider having the job professionally done. The process consists of first smoothing the floor as described above, then applying a liquid adhesive-stripping product, followed by sanding with a floor polisher equipped with an abrasive pad. The floor must be kept wet continuously during the process by spraying it with detergent solution. Areas the polisher cannot reach must be sanded by hand. Proper cleanup involves vacuuming the slurry produced by sanding with a HEPA wet-dry vacuum, rinsing the floor with plain water, and then vacuuming again after the floor has dried.

Tiles, even loosened ones, installed over a wooden floor seldom need removing, as new underlayment generally can be installed over them to accommodate new tiles. Even if tiles must be removed, installing new underlayment eliminates the need for smoothing and removing adhesive. However, if tile flooring containing thin underlayment must be removed, this can be done by prying up the underlayment panels as described for sheet vinyl flooring. Getting rid of old tiles by this method reduces the number of tiles that must be removed individually and often leaves a bare floor ready for resurfacing with little preparation.

Lead

What Is It?

LEAD IS AN ELEMENT, ONE OF THE BASIC SUBSTANCES OF THE earth's crust. Its chemical symbol is Pb, from the Latin word *plumbum*, meaning "waterworks." Ancient Roman water pipes were often made of lead, hence the derivation of the term. Lead is everywhere in the human environment as a result of industrialization. Currently, the United States is the world's largest producer and consumer of lead.

Lead's atomic number, its position on the Periodic Table of the Elements, is 82. Its atomic weight is 207.20, making it the heaviest of the common metals, including gold. In its most familiar form, lead is a lustrous, silvery metal—very soft—that tarnishes quickly in contact with air and becomes a dull, bluish gray. Besides its softness, one of lead's chief properties is that it melts at a very low temperature, 662 degrees Fahrenheit. This is easy to achieve without elaborate heating methods; in fact, while it is dangerous to do so because of poisonous fumes, lead can be melted over a wood fire or in a frying pan on an ordinary kitchen range.

Civilizations older than the Romans knew of lead. The metal is mentioned in the Old Testament, and the Egyptians used it in coins, weights, ceramic glazes, utensils, and solder. Later, in the eleventh century A.D., the Venetians added lead in the form of lead monoxide (a yellow crystalline powder commonly called litharge) to glass, which made the glass easier to work with and enabled them to produce the thin sheets and intricate shapes for which Venetian glass was famous. In the seventeenth century, English glassmakers added even more lead (amounting to nearly 30 percent of the total composi-

26

tion) and developed glassware known as leaded glass or crystal. This type of glass is still manufactured and continues to be prized now, as then, for its transparency, sparkle, and other optical qualities.

Paint was also discovered to benefit from the addition of lead compounds, principally lead carbonate, which is commonly called white lead. Chrome yellow, a brilliant hue, and its relatives—chrome red, orange chrome yellow, and lemon chrome yellow, among others—actually contains lead chromate; red lead, or minium, used on bridges and for marine bottom paint, contains lead oxide. Paint containing lead is more durable than other varieties, and it sheds or chalks readily, thereby continually exposing a fresh, clean surface that reduces the need for repainting.

Today, besides the properties that have made lead valuable throughout history, it is in demand chiefly for its shielding abilities. It is virtually impervious to corrosion and does not conduct electricity. Its density and softness enable it to dampen sound and other vibrations (heavy machinery and even buildings are sometimes placed on lead blocks to isolate them or their surroundings), and it provides the most efficient protection against powerful atomic radiation. Common contemporary uses of lead are battery plates, tank linings, plumbing, shielding for electrical wires, X-ray facilities and other radioactive installations, and, as mentioned, in chemical compounds like paint. In addition, tetraethyl lead—an organic compound consisting of lead, carbon, and hydrogen—has until recently been used in gasoline as an antiknock additive.

Of course, lead is also a well-known toxin. As he did with asbestos (see page 4), Pliny the Elder noted the ill effects of lead on Roman slaves extracting and working with the metal during the first century A.D.; and since the eighteenth century various symptoms have been attributed to lead exposure, among them violent intestinal pain once characteristic among housepainters and others in the paint trades. Referred to then simply as "painter's colic," today the condition is recognized as a characteristic sign of severe lead poisoning.

Lead usually enters the bloodstream when it is swallowed as dust, paint chips, or as particles in contaminated drinking water. Lead can also be inhaled, either as dust or as fumes from molten lead. After circulating through the blood vessels, most lead accumulates in the bones and teeth (severe lead poisoning can produce a black line along the gums), and to a lesser degree, in the kidneys and intestine.

While a massive single dose of lead—for example, a mouthful of lead-based paint chips—can cause symptoms requiring emergency treatment immediately, lead typically enters the body in small, frequent doses that produce no visible signs until high levels are reached. Obviously, such cases of lead poisoning frequently go undetected; unfortunately, even when they are discovered, damage has usually occurred that may be irreversible.

Severe lead poisoning, which occurs when levels of lead in the blood exceed 60 micrograms per deciliter (ug/dL) in children and around 150 ug/dL in adults, can produce irritability, headaches, hallucinations, dullness, stomach aches, hypertension, kidney damage, anemia, nerve damage, convulsions, coma, paralysis, and death. In children, physical and mental retardation and encephalopathy (inflammation of the brain) are also characteristic.

Moderate lead poisoning—which occurs in children when blood-lead levels are between 35 and 50 ug/dL and in adults when levels exceed 70 ug/dL—can slow nerve impulses, cause fatigue, short-term memory loss, reduced hand-eye coordination, increased blood pressure, and stroke; lead to low sperm count, impotence, and sterility in men; and increase chances of miscarriage, stillbirth, and premature delivery in women.

Even the lowest blood-lead levels at which children are now considered at risk—10 ug/dL—have been associated with reduced IQ, learning disabilities, hyperactivity, hearing impairment, kidney abnormalities, and slowed physical and mental growth. The latter apparently is due to disturbances in the formation and maintenance of red blood cells, impaired ability to metabolize vitamin D, inability to absorb iron and use calcium, and disruption in the creation and function of important enzymes and amino acids.

Children, especially those under six years of age, are at the greatest risk from exposure to lead. Children's breathing and metabolism rates are faster than those of adults and so result in higher intake of lead and greater absorption of the metal through the stomach lining and lungs. This happens at a time when their developing bodies, including the brain and central nervous system, are most susceptible to harm. Young children also are more likely to ingest lead than older people because they are more often exposed to its chief source—contaminated dust and soil—at an age when frequent hand-to-mouth activity is normal. Finally, less lead is required to induce poisoning in children than in adults because their bodies and blood volumes are smaller.

Pregnant women face almost as much risk, although actually it is the fetus that is most vulnerable. Lead passes easily across the placenta, and children have been born with levels of lead in their bloodstreams that were 95 percent as high as those of the mothers. At high levels lead can cause prenatal damage to organs and organ systems; and at any level lead absorbed before birth may cause mental or physical retardation after a child is born.

Other adults generally do not suffer as much, principally because their bodies, brains, and central nervous systems are fully developed and so are not as susceptible to lead's stunting effects on growth. Middle-aged men can develop high blood pressure from lead intake, besides the difficulties mentioned earlier. Among both sexes, smokers and people already with high blood pressure or heart disease are at greater risk than nonsmokers and persons without these diseases. Among the elderly, many symptoms often attributed to normal aging—muscle pain, joint aches, and memory loss—actually may be undiagnosed signs of lead poisoning.

How Bad Is the Problem?

Lead has no known function in the body, and while information about its effects at blood-lead levels lower than 10 ug/dL is sketchy, there is no recognized minimum exposure level below which effects are known not to occur.

Since the 1970s, when the federal government began focusing attention on lead poisoning through the newly organized Environmental Protection Agency (EPA), research findings have caused thresholds for taking medical or other action against lead poisoning to be set at lower and lower levels. The current level of 10 ug/dL, set by the Centers for Disease Control (CDC) in 1991, is down from 25 ug/dL, set in 1985 (in 1970, the CDC's threshold was 40 ug/dL!).

On the plus side, blood-lead levels in children have plummeted during the past twenty-five years. The CDC reports that deaths and severe lead poisoning are rare; in 1991, the EPA reported that the percentage of children with blood-lead levels higher than 25 ug/dL dropped from 10.7 percent to 1 percent between 1976 and 1990 and that the percentage of children with levels higher than 10 ug/dL dropped from an astounding 91 percent to only 15 percent. Much of the decline is attributed to the phasing out of lead in gasoline (complete elimination took effect December 31, 1995, as part of the federal Clean Air Act). During the past fifteen years or so, lead also has been removed from plumbing solder, drinking fountains, and food cans, all of which once were major sources.

On the minus side, that same year the EPA estimated that one child in ten, including over half of all children in many large cities, had blood-lead levels exceeding the current action threshold. But does this mean that millions of American children are lead-poisoned? No. The CDC does not refer to children with blood-levels below 14 ug/dL as "poisoned," nor does it suggest medical evaluation even for children with levels under 20 ug/dL. The 10 ug/dL action threshold, though significant, is intended primarily to signal communities that a concentration of children having blood-lead levels higher than this amount may indicate the presence of a local enironmental hazard requiring investigation.

LEAD BLOOD-LEVEL ACTION THRESHOLDS AND FOLLOW-UP ACTIVITIES FOR CHILDREN (Adapted from CDC)

Blood-level Concentration	Follow-up activity Recommended by CDC
Below 9 ug/dL	A child with this blood-lead level is not considered to be lead-poisoned.
10–14 ug/dL	A large proportion of children in a community with blood-lead levels in this range should trigger community-wide childhood lead poisoning prevention activities. Children in this range may need their blood-lead levels tested more frequently.
15–19 ug/dL	A child in this category should receive nutritional and educational interventions and more frequent screening. If the blood-

lead level persists in this range, environmental investigation should be done.

20–44 ug/dL

A child with blood-lead levels in this range should receive environmental and medical evaluation. Such a child may need pharmacologic treatment for lead poisoning.

45–69 ug/dL

A child with blood-lead levels this high will need both medical and environmental interventions, including chelation therapy requiring hospitalization. (Chelation includes the administration of E.D.T.A.— calcium disodium ethylenediamine- tetraacidic acid. Lead is removed from the body by displacing the calcium in E.D.T.A.; the resulting stable compound then is excreted in the urine.)

70 ug/dL or higher

A child in this condition is a medical emergency. Medical and environmental intervention must begin immediately.

Dust and soil containing lead from paint is considered the most common contemporary source of lead poisoning, according to the CDC, especially in children. The Department of Housing and Urban Development (HUD) estimates that at least 74 percent of all privately owned, currently occupied houses and apartments (about 57 million units) built in America before 1980 contain some lead. Until the 1950s, housepaint containing as much as 50 percent lead was in wide- spread use; and while levels of lead in paint declined considerably during the next twenty years, it was not until 1978 that the Consumer Product Safety Commission (CSPC) banned the manufacture of paint containing more than 0.06 percent lead by weight for use on residen- tial surfaces, toys, and furniture. Larger amounts of lead continue to be allowed in paint for military, industrial, and marine uses, however; and this paint occasionally finds its way into homes despite the ban.

The paint that contributes lead to dust typically is that which has flaked or chalked due to aging, or has been disturbed during home maintenance or renovation. Indoors, lead paint generally is found on kitchen and bathroom walls, and in older houses on doors, windows, and trim. Naturally, paint or painted surfaces in poor condition pose a greater risk than those that are sound, but even intact paint can be a hazard if located on projecting surfaces like windowsills that can be mouthed or chewed. CDC researchers believe fewer children eat paint chips than eat or inhale contaminated dust and soil; of particular concern is lead paint on window frames, because it is regularly ground to dust by the repeated opening and closing of the window sashes.

Outdoors, lead has been found to accumulate in the soil around the perimeter of older houses that have undergone repainting. Some areas have been found with lead levels higher than ten thousand parts per million, according to the EPA, who, together with the Agency for Toxic Substances and Disease Registry (ATSDR), estimates that blood-lead levels typically rise 3 to 7 ug/dL for every thousand parts per million increase in soil or dust lead concentrations.

Other sources can also contribute lead to soil. Because lead does not dissipate, biodegrade, or decay, all that has been deposited by exhaust emissions during the years when gasoline contained lead is still present. According to the ATSDR, this amounts to an estimated four to five million metric tons, much of it within the top two to five centimeters (approximately half an inch) of ground that is undisturbed.

On the other hand, lead in household drinking water is not considered a major source for poisoning nationally, according to the CDC. Lead levels are usually low in groundwater or surface water, but they can increase when the water enters a distribution system containing lead plumbing and/or lead solder. Lead plumbing typically is found only in homes and other buildings built during or before the 1920s. Copper plumbing, joined with solder containing lead, came into general use during the 1950s and was not banned until 1988; even risk from these systems is comparatively slight, however, because ordinary water typically contains mineral salts that form a coating on pipes, covering soldered joints and also slowing the formation of soluble lead hydroxide. Areas with so-called hard water—water containing elevated amounts of minerals—are especially protected,

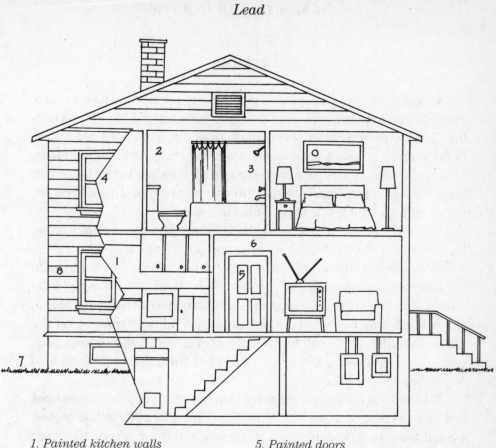

1. *Painted kitchen walls*
2. *Painted bathroom walls*
3. *Faucets and shower*
4. *Painted window trim*

5. *Painted doors*
6. *Painted molding and other trim*
7. *Soil around house exterior*
8. *Painted exterior siding*

Illustration 8. Sources of household lead

because coatings inside pipes form quickly and attain considerable thickness. In fact, if you live in such an area and your plumbing contains lead pipes or lead-soldered joints, think twice before installing a water softener; doing so may strip the pipes of their protective deposits.

Nevertheless, lead concentrations in household water can vary considerably, even between homes in the same neighborhood. The only sure way to determine the condition of your water is by having it tested by a laboratory, as described further on in this chapter. Testing is strongly recommended if a member of the household is pregnant or if young children drink the water.

Testing

Whether or not you believe your house presents a lead hazard, many health experts, including the CDC, recommend blood-lead testing for all children under seven years of age at least once before the child is three years old and again when the child is six (some areas, notably New York City, require testing of all children under the age of two). Older children and adults who believe they may have been exposed to high lead levels should also be tested.

The most accurate testing method is the direct-lead test. It must be administered by a doctor or clinic and is performed by sticking the finger or drawing blood from a vein. A frightened child or inadequate facilities may necessitate the finger-stick method, but drawing blood from a vein is usually more accurate because there is less chance that the sample may become contaminated by lead dust present on the skin. An older test—called an EP, or erythrocyte-protoporphyrin, test—is considered unreliable by the CDC for indicating blood-lead levels below 25 ug/dL.

If blood-lead testing is negative you can assume your house does not pose a hazard, at least in its present condition. As with asbestos, even if the house contains lead, if the material is not posing a current problem and there are no plans to disturb it, leaving it alone is the best and least expensive course of action. However, if you plan to renovate, which can possibly create a lead hazard, you should have the paint tested beforehand on any areas likely to undergo work.

To have paint in your home tested, consider performing initial testing yourself with an inexpensive lead-testing kit. There are several kinds, but all contain either rhodizonate, a chemical that turns pink or red in contact with lead, or sodium sulfide, which in contact with lead turns gray or black. Rhodizonate is commonly used in criminal investigations to determine the presence of lead from bullets.

Both kinds of testing kits are available at many hardware stores and home centers; other sources are paint stores, safety products suppliers, and test kit manufacturers listed in the section "Where to Find Help." Prices for kits range from about $5 to about $30. Most kits contain materials for about five individual tests.

To get an accurate picture of an entire home's potential or existing lead hazards, every surface that may contain lead paint should be

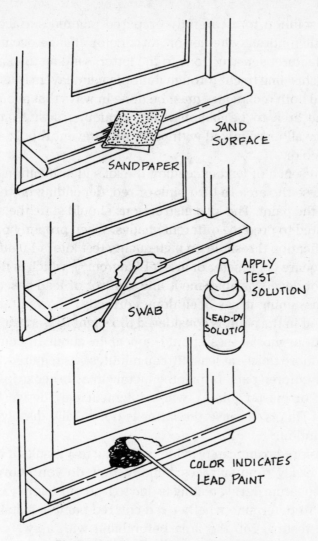

Illustration 9. Testing paint for lead

tested. However, the CDC points out that extrapolating results when they are positive can reduce the need for overly extensive testing. For example, if painted trim in one room tests positive for lead you can assume that similar trim in other rooms contains lead also. To remain on the safe side, do not extrapolate when results are negative. In other words, should testing indicate no lead is present in one location, do not assume that other, similar locations are also free of lead.

In general, the testing procedure involves applying a few drops of chemical solution to a carefully prepared painted surface or paint chip and then noting whether or not a color change occurs. Follow the manufacturer's instructions to the letter; sanding to expose fresh paint and chiseling to cut through multiple paint layers is usually necessary, and both techniques must be done in ways that avoid contaminating the area to be sampled. For example, dust from gypsum plaster or wallboard mixed with paint will prevent rhodizonate from changing color.

The presence of lead exceeding the kit's identification threshold should cause the area to turn pink or red, depending on the amount of lead in the paint. The kit's instructions should state the amount of lead required to produce different shades. When properly performed, the identification threshold on high-quality test kits is 1.0 milligram of lead per square centimeter of area ($1.0mg/cm^2$), which is the current HUD action level for abatement and control of lead-based paint. If the test (assuming it has been done correctly) shows no color, any lead present in the paint is considered too slight to warrant attention.

Results in most cases should appear in about fifteen seconds. However, more time—at least fifteen minutes—is required when testing yellow-colored paint or painted metal (like corner strips covering wallboard or plaster joints), whose pigment may derive from lead chromate. This is because the latter is not readily dissolved by the testing solution.

Red-colored paint can provide confusing test results if it is a type that rubs easily from surfaces. Chiefly, how do you know whether any red appearing during testing is due to lead or is simply rubbed-off pigment? To determine whether red-colored paint is suitable for rhodizonate testing, rub the area beforehand with a swab dipped in plain vinegar or in a small amount of the kit's leaching solution, if there is one. Use only a swab included with the kit; these are specially formulated to be compatible with the testing materials. If the swab turns red, test the paint with a sodium-sulfide kit instead (these produce excellent results in any case), or choose one of the other testing methods described further on.

For testing soil, send samples to a laboratory recommended by your county agricultural extension office (listed in the phone book). While rhodizonate kits technically can be used to test for lead in soil, their results vary with soils of differing types (clay soil versus sandy soil, for example).

If spot tests performed with rhodizonate or sodium sulfide are positive, you know that lead levels on the surfaces tested are at or above the lowest level of lead currently allowable in residential housepaint. Should this occur, call your local health department, state agency for lead, or regional office of the EPA for advice before proceeding.

Likely, further testing by professionals will be prescribed both to confirm the results of your own tests and to determine the actual amount of lead present. One testing method—sending samples to a laboratory for analysis—you can do yourself with care. (Mail-in testing kits are available from sources listed in "Where to Find Help.") The other methods consist of scanning surfaces with special instruments whose operation requires advanced training and experience.

Laboratory testing scores highest for accuracy among the various methods, provided sampling is done correctly to avoid contamination. Generally, sampling involves chiseling a small area of paint from a surface after covering the area with clear plastic adhesive tape. Practice is recommended—and you should be aware that the technique creates unsightly areas that must be patched. For a list of recommended lead-testing laboratories contact your local health department, state agency for lead, or a regional office of the EPA or OSHA. A list is also available from the American Council of Independent Laboratories, 1629 K Street N.W., Suite 400, Washington, D.C. 20006. The cost for sending paint samples to a lab for analysis generally ranges between twenty and thirty dollars per sample. Results typically arrive in one to three weeks.

Contact the lab in advance to obtain any special instructions on gathering samples to its specifications. Ask also to receive the results in mass per unit area (mg/cm^2); then you can easily compare it to the HUD action level, which is in the same terms.

Professional instrument testing is usually done with a device called an X-ray fluorescence analyzer, or XRF. No sampling is required and test results usually are available immediately (actually within five to ten seconds) by reading them from the instrument. The drawbacks of XRF analysis are that readings cannot be taken of curved surfaces or where access is awkward (the device is cumbersome). Also important, readings are not highly accurate at low concentrations of lead like those near the HUD action level. However, testing large numbers of household surfaces—important when plan-

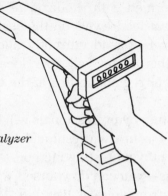

Illustration 10.
An X-ray fluorescence analyzer

ning a full-scale remodel—is quite feasible using an XRF, and unlike the physical testing methods described earlier, an XRF can disclose lead in subsurface paint layers without having to sand or cut into them. Usually, the cost for XRF analysis of a single-family house runs from $300 to $500. (Costs for having a similar number of individual samples evaluated by a laboratory usually would be far higher.)

In 1994, another instrument capable of onsite testing, the anodic stripping voltameter, was introduced. While not yet as widely available as XRF devices, these solid-state devices cost less to manufacture, resulting in lower testing costs (about $200 for a house), take readings more easily than XRFs, and are more accurate than the latter at low lead levels.

Lead-testing companies are listed in the telephone book. To select a qualified company, contact your local health department or one of the state or federal agencies mentioned above. If a list is not available, ask what qualifications lead-testing companies are required to possess in your area and then, as you would when selecting any contractor, interview several testing companies before making a selection.

Getting Rid of Lead

Professional removal or abatement of lead, while recommended as the least hazardous solution to the problem, can be prohibitively expensive. Estimates published by HUD in 1990 placed the average cost for deleading a single-family home at between $5,500 and $11,900. For some 15 percent of the houses surveyed estimates exceeded $25,000.

If anything, costs for deleading have risen along with other costs of remodeling. However, the fact is that by taking precautions, knowing how to proceed, and working carefully, amateurs and untrained professionals often can control, reduce, and even eliminate lead hazards in homes for a good deal less money than the HUD estimates suggest. Be warned that doing a poor job of deleading can make a situation far worse and can even create a hazard where previously none existed. Both the EPA and the Consumer Product Safety Commission (CPSC) recommend professional removal of lead from homes. If you decide to hire a contractor, find one who has completed a course at one of the EPA's six regional lead-abatement training centers. Be sure the contractor plans to follow the procedures outlined in this chapter, and specify that the job must include follow-up testing by someone other than the contractor who performed the work. At present, twenty-two states license or certify lead-abatement specialists. That said, the following are the procedures currently recommended by relevant organizations, including the CDC, EPA, OSHA, and the National Association of Home Builders (NAHB).

As mentioned earlier, lead-bearing surfaces in sound condition and not likely to be mouthed or chewed by children or pets generally can be left alone. However, dust from such surfaces must not be allowed to accumulate; this means floors and other horizontal surfaces must be damp- or wet-mopped at least weekly using a strong detergent mixed with water. (Any heavy-duty detergent will do—powdered dishwashing detergent is recommended; originally, the CDC and other authorities recommended using only high-phosphate detergents for cleaning lead-contaminated areas, but tests since have shown that ordinary heavy-duty detergents with little or no phosphates are virtually as effective.) When washing surfaces, wring out the mop water into a bucket, and use a separate bucket containing clean water for rinsing; that way you will not simply reapply lead dust to the floor.

Reserve one set of cleaning materials—sponges, rags, and mops—for cleaning leaded surfaces only. Especially avoid using these materials to wash dishes or wipe eating surfaces like tables and countertops.

Keep cribs and playpens away from even sound lead-painted surfaces, particularly window and door trim. Avoid sweeping or vacuuming hard floors, as these activities merely spread dust and lift it into the air. Carpeting, drapes, and upholstered furniture, all of which can harbor vast amounts of dust, are not recommended in homes containing lead and should be discarded. If you choose not to remove carpeting, vacuum it often using either a HEPA vacuum or a vacuum cleaner equipped with a powered rug-cleaning attachment. Select an attachment with a rubber gasket around the base to further contain dust. Don't be tempted to use a fresh vacuum cleaner bag; in fact, it is crucial that the bag be previously used and therefore partly clogged with dust. While a fresh bag will not contain paint dust—the vacuum will simply spew it around the room through the exhaust—evidence has shown that a partly clogged bag retains dust particles whose size is nearly as small as those trapped by a HEPA vacuum.

Attend diligently to children living in houses containing lead paint. Wash their hands often, especially before eating; wash pacifiers and chewable toys daily. Interestingly, diet can reduce lead hazards; eating foods rich in iron (spinach, liver, fortified cereal) and calcium (milk, cheese, cooked green vegetables) reduces the body's ability to absorb lead, according to the CDC.

If your home contains chipping or peeling lead-based paint you must take some action. At least place furniture in front of problem areas to make them less accessible and then follow the maintenance procedures described above. As with friable asbestos (see the chapter "Asbestos"), loose paint inevitably must be removed, encapsulated (sealed with special paint or another substance), or enclosed by covering it with rigid material like wallboard or siding. All three methods are discussed further on; which method you choose depends largely on cost and the effort required. Crucial to all operations are working in a way that minimizes the creation and release of new lead-bearing dust, and cleanup that removes lead-bearing dust entirely.

To prepare for work involving only a small area—for example, removing or refinishing a window frame—it is usually enough to move nearby furnishings and then cover the floor with six-mil polyethylene sheeting that extends at least four feet beyond the area in all directions.

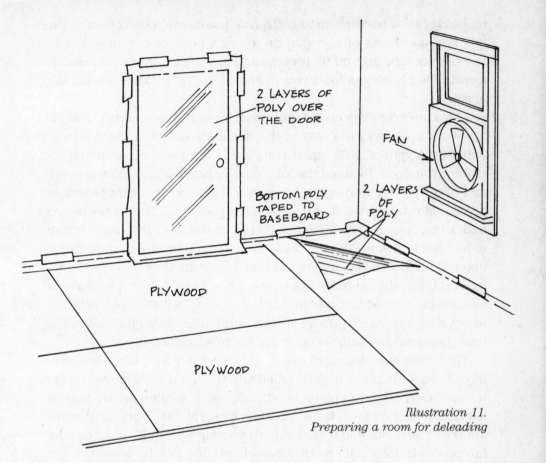

2 LAYERS OF POLY OVER THE DOOR

FAN

BOTTOM POLY TAPED TO BASEBOARD

2 LAYERS OF POLY

PLYWOOD

PLYWOOD

Illustration 11.
Preparing a room for deleading

For larger areas—say, several windows or a wall—you should take pains to remove furniture, drapes, carpet, and any other portable items to protect them from dust. Cover with plastic sheeting anything that cannot be moved.

Clean the floor if it is dusty and then cover it entirely with at least one layer of six-mil polyethylene sheeting taped to the baseboard to prevent new dust from creeping underneath. Some remodelers set down two layers when work will likely proceed for several days; at the end of each day the top layer then can be easily bundled for disposal with paint chips and dust inside.

Another technique is to cover the floor with overlapping strips of poly that are each two or three feet wide. This secures the material at several places across the floor—not just around the edges—which makes it easier to work on and clean. When demolition is part of the job, arrange sheets of plywood on top of a layer or two of poly and fasten duct tape around their edges to keep dust from getting underneath. The plywood allows you to shovel debris without tearing the plastic.

Also on large projects (and on small ones if appropriate), seal off the work area from the rest of the house by covering the doorway with two layers of poly taped completely around the perimeter, including the floor. To avoid tracking dust or causing it to blow into the rest of the house, plan not to leave the area between cleanups unless you can do so by a window. If this is not possible, create a protected path to an outside door by placing poly on the floor leading to it and either erecting walls of poly if the area is open or else sealing extraneous doorways if the path is in a hall. Occupants not involved in de-leading tasks should leave the house when work is being performed and ideally should not return until after the final cleanup. Alternatively, they should stay away at least until after daily cleanups. Children, pregnant women, and pets are the most vulnerable.

To ventilate sealed work areas, use a portable fan placed in a window. Unless the room is quite small and the fan is quite powerful (in which case it should not be used), only an insignificant amount of lead dust will be blown to the outside. General dust from large work areas can be contained by attaching a heating system filter across the fan on the indoor side; in all cases check that nearby windows (including those of neighbors) are kept closed when the fan is operating, and avoid aiming the fan where passersby might pass through the stream.

Wear protective gear and clothing while working, even on small projects. Proper gear includes a dust mask or half-face cartridge respirator approved by the National Institute of Safety and Health (NIOSH) for protection against lead dust and fumes, safety glasses or goggles, and work gloves. Clothing should consist of shoes, long pants, a long-sleeve shirt, and a painter's hat if practical. Make arrangements for changing into clean clothes before entering the rest of the house or traveling to a different location. Transport work clothes in a plastic trash bag; launder them separately from other clothing. Professionals often wear disposable suits made of breathable synthetic material like Tyvek. These are available from safety equipment suppliers; check the telephone book for sources.

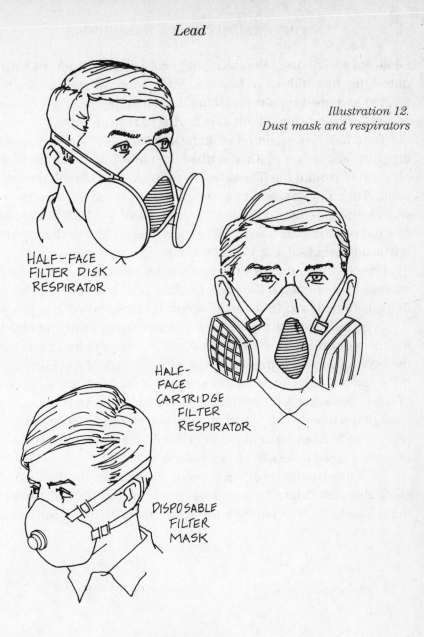

Illustration 12.
Dust mask and respirators

HALF-FACE
FILTER DISK
RESPIRATOR

HALF-
FACE
CARTRIDGE
FILTER
RESPIRATOR

DISPOSABLE
FILTER
MASK

Removal

The surest abatement procedure is removing lead-painted components intact, either for replacement with new material or for refinishing off the premises. However, it is important to produce as little dust as possible during the process and to clean up meticulously afterward. Usually, removal is most feasible with molding and trim; where finances allow, it is also the best way to cope with lead-painted windows and doors. (Replacing old lead-painted windows with new ones often has additional benefits. Such windows usually are not insulated and need repair; replacing them increases overall household energy efficiency and eliminates having to restore them.)

To remove lead-painted molding and trim, first wash off any peeling paint with a wet cloth and rinse the chips into a bucket of water. Next, slice around the pieces with a utility knife to break through any paint. Then pry gently along the entire length of each piece with a wide, stiff-bladed putty knife or scraper until you can insert the end of a flat pry bar into the gap. Work gradually; old woodwork often is brittle and breaks easily if bent too far.

Protect the wall from the pry bar by placing the putty knife or scraper blade beneath the bar's fulcrum. The bar will be most effective positioned next to the nails securing the woodwork to the walls.

Once the pieces are pried away, remove nails from them by drawing them out through the rear with pliers. Channel-locking pliers are the best tool to use; grasp the nail with the tips of the jaws close to the wood and then lever out the nail by rocking the curved upper jaw of the pliers against the surface. This technique avoids damaging the front of the woodwork, which mars it and also creates paint chips. To remove nails from walls, use a pry bar or a claw-hammer with a block of wood placed beneath it to increase leverage.

To remove windows, wash them thoroughly to eliminate paint chips and dust from all corners and crevices; pay special attention to the sill and the area beneath the lower sash. Remove the sashes to

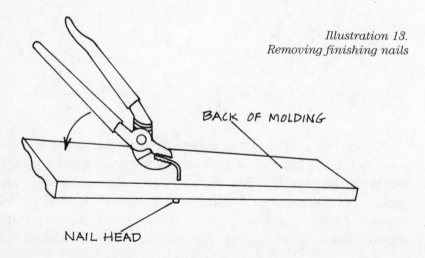

Illustration 13.
Removing finishing nails

BACK OF MOLDING

NAIL HEAD

avoid breaking the glass and pry away the trim as described above. Then wash the frame again, including the area above the top. Free the window frame from the wall by sawing through the nails in the jambs (sides) holding it in place or by sawing through the jambs and prying. Collect debris in a bucket or on plastic sheeting spread beneath the window. Pass the removed pieces directly to the outside.

Consider having woodwork professionally stripped if it is to be refinished. That way you need not have anything to do with this messy task and its potential risks. If you do decide to tackle the job, work outdoors or in an open garage. Stripping woodwork with paint remover produces the least lead dust but is the most expensive method besides professional stripping.

The safest paint removers to use are those containing dibasic esters (DBE), n-methyl-2-pyrrolidone (NMP), and gamma butyrolactone (BLO). These usually release no fumes, are seldom flammable, and irritate only the most sensitive skin. Of the three, paint remover containing DBE is the mildest—but it's also the slowest acting. The other two types work more quickly but are somewhat more irritating to skin. Follow all product directions and safety precautions carefully. In particular, be sure to coat pieces thickly with remover and allow it to remain on the wood for the required amount of time.

Scraping and sanding are feasible if materials are kept wet during the process. Mist the woodwork with a spray bottle containing plain water; scrape with a scraper or sand using a sanding sponge or wallboard-sanding screen. The trick is to apply enough water to prevent dust from becoming airborne but not so much that you create thick sludge. Work over newspapers placed on plastic sheeting. To clean up, bundle the newspapers to contain the waste and discard them with the trash.

Whether scraping or sanding, do not attempt to remove sound paint. When finished with a piece, dry it promptly and then fill the cratered areas with spackling compound to level them with the surrounding surface. When the spackling hardens you can dry-sand it lightly to smooth it (spackling dust contains no lead and the amount abraded from the sound paint will be negligible), or try rubbing it instead with a damp cloth or sponge. The latter method requires some practice but produces no dust.

Rinse sanding materials often in a bucket of water as you use them, to remove chips and dust. Later, strain the bucket's contents through a cloth to remove the paint chips for disposal. Flush the dirty water down a toilet.

Paint remover, wet-scraping, and wet-sanding can be used to strip woodwork even if you cannot, or decide not to, remove it. To avoid a mess, blot walls before water and residue drip to the floor.

Or consider using one of the poultice paint removers. These incorporate a canvas covering or contain a thickening agent that develops a skin, which slows evaporation of the removal chemical. When the paint has dissolved, the skin, together with the remover and paint adhering to it, then can be pulled away from the surface and disposed of by placing the bundled mass in a plastic bag. Poultice paint removers are more expensive than the ordinary liquid or gel varieties

Illustration 14.
Poultice paint remover

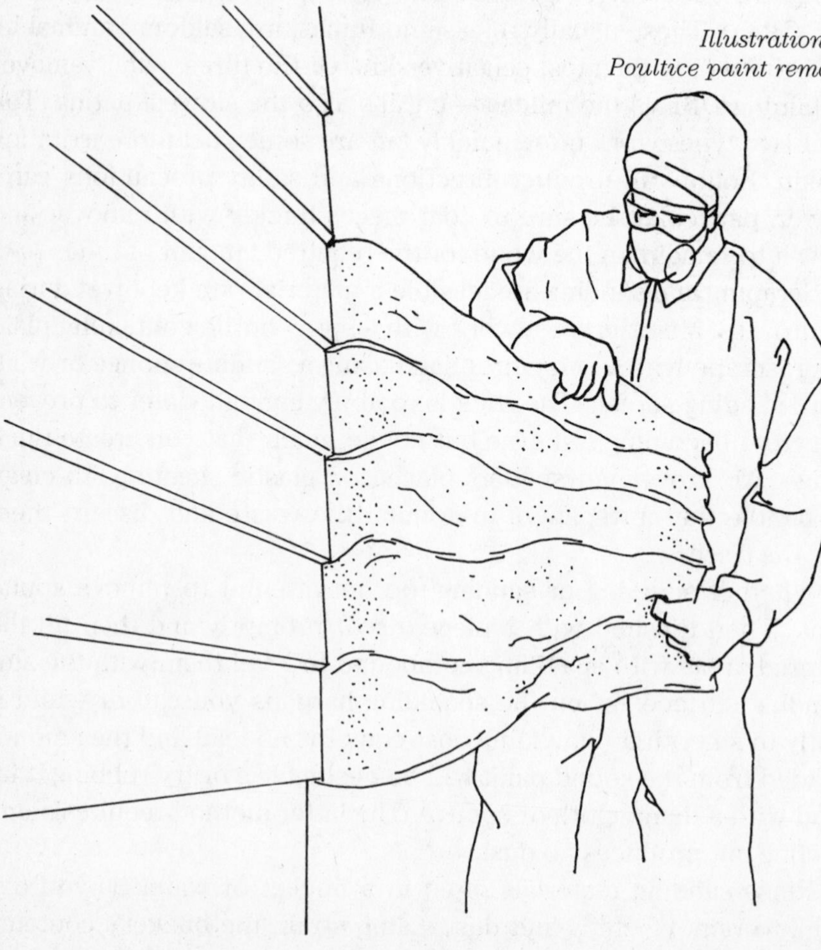

but generally are more effective at removing thick layers of paint and are easier to use for stripping large surfaces like masonry walls. Alkaline, solvent-based, and acid-based formulas are available, each having specific uses and requiring specific safety precautions. You can obtain poultice paint removers through paint supply stores. Consult the manufacturer's recommendations and instructions before ordering.

Most other paint-removal methods, indoors or out, are riskier and not recommended. Ordinary dry-scraping and sanding usually produce clouds of lead-bearing airborne chips and dust. Using a heat gun to soften paint for scraping can vaporize lead, creating toxic fumes; and softening or burning paint with a blowtorch not only creates dangerous fumes but is an extremely risky fire hazard. Leave the removal of large areas of paint from outdoor surfaces to professionals; in a growing number of areas, deleading of outdoor surfaces must be performed by a licensed professional abatement contractor in any case.

After stripping woodwork by whatever method you choose, wipe it free of residue with mineral spirits and then coat it with high-quality primer before repainting it. For repainting, high-quality acrylic latex paint is recommended because it is water-based and because cleanup is easy, requiring only soap and water. Alkyd-based enamels are also acceptable (they sometimes may be marginally more durable in areas like kitchens and bathrooms), but these are solvent-based and can produce irritating fumes during application and until the paint dries completely. Cleaning up after applying alkyd-based paint requires using paint thinner, which is messy and also emits irritating fumes.

Encapsulation and Enclosure

Sealing, or encapsulating, lead-painted surfaces, or enclosing them, often creates less dust than does removal and generally is much less expensive. The chief drawback is that neither method removes lead from the premises; it merely prevents contact with it. This can become a problem later if the encapsulating or enclosing material wears away or if remodeling involving demolition is done that exposes the lead. Federal law now requires landlords and sellers to disclose known information on lead-based paint hazards in apartments and homes built before 1978, and to allow prospective tenants and buyers up to ten days to check for lead hazards.

To encapsulate lead-painted surfaces, apply a paintlike elastomeric coating or a veneer wallcovering material like spun fiberglass. The former is applied in two layers using a brush, roller, or sprayer; the latter is applied like wallpaper, usually with paint as the adhesive. Both materials are described more completely in the chapter on asbestos. Elastomeric coatings are available at some paint stores and from safety equipment suppliers; veneer wallcoverings can be obtained through building supply stores, wallcovering stores, and interior decorators. Ordinary housepaint is not recognized by most authorities as an effective encapsulant.

To enclose surfaces, simply cover them with wallboard or other rigid paneling. Outdoors, new siding usually is the answer.

In all cases after deleading a house or room, test again in several places to make sure results are satisfactory and that cleanup has been sufficient.

Removing Lead from Water

EPA actions levels for lead in drinking water are fifteen parts per billion (ppb) from first-draw water (water sampled after standing in pipes for several hours or overnight) and five parts per billion from purged-line water (water sampled from a faucet that has been allowed to run a minute or more). If your water tests higher than these amounts, you should contact your local health department.

Steps you can take to reduce the amount of lead in water are to let water run for at least a minute from faucets before using it and to always draw water for drinking and cooking from cold water supply lines, never from hot water lines.

Adding filtration equipment to water lines can help, but is likely to be expensive and may have dubious results. According to the CDC, two systems are most effective: reverse-osmosis units and distillers. The former require professional installation, but generally are regarded as the best at removing lead, which they do by pumping water through a semipermeable membrane filter that screens out most molecules larger than water. Drawbacks to reverse-osmosis systems are their expense—usually $500 to $800 for an undersink model, and

$300 to $500 for a countertop model—and their slow performance, which may not meet the daily drinking water demands of a large family. Filters for reverse-osmosis units can be easily clogged by hard water minerals and require periodic replacement in any case, at a cost ranging from about $50 to over $200, depending on design (fortunately, the more expensive filters require replacement less often than the cheaper ones).

Distillers simply boil water and then collect the steam as it cools and condenses. The process leaves behind all materials that do not boil readily—namely lead and other minerals—and so deliver water that is pure in this respect. While not requiring professional installation and not as expensive as reverse-osmosis devices (prices range from about $200 to about $450), distillers operate very slowly—often taking more than an hour to produce a quart of purified water—consume large amounts of electricity, and clog easily with mineral deposits.

The jury is still out on a third system—filters that can be installed by do-it-yourselfers. While relatively inexpensive (average cost is around $100) and touted by advertisers as both effective at "purifying" water and easy to install, the truth is that few are very good at removing lead. Those that work best contain cartridges of activated alumina, and even these are not highly regarded by the CDC, although tests conducted by *Consumer Reports* show good results. Filters containing only carbon are not effective at removing lead.

Before purchasing any unit, ask to see results of independent testing and find out your rights should the unit you buy not perform as advertised.

Radon

What Is It?

RADON IS A GAS FORMED BY THE NATURAL BREAKDOWN, OR
decay, of uranium in the earth's soil. It is colorless, odorless, and
tasteless; unfortunately, it is also radioactive. Virtually all soil contains some uranium and therefore some radon. Higher-than-average
amounts of uranium are found in soils containing granite, phosphate
rock, and a type of shale called black shale.

Radon rises through the soil and also is carried by groundwater.
Outdoors, when radon reaches the surface it poses little danger to
health because it mixes almost instantly with enough surrounding air
to effectively dilute and disperse it. But when radon rises or is carried
into an enclosed space like a house, less air is available. This means
that the gas is neither diluted as much nor as quickly dispersed. Both
conditions result in increased levels of radiation that are potentially
harmful to health.

Actually, the main source of radioactivity associated with radon
is not the gas but the series of elements formed when it, too, decays.
These elements, called radon daughters or radon progeny, are polo-
nium-218, lead-214, bismuth-214, and polonium-214. The numbers indicate the elements' atomic weights. (Each element lasts only a few
moments—polonium-214 lasts barely a fraction of a second—before
changing to the next element in line, and finally to lead-210, which is
relatively stable.) All of these substances are ultrafine particles—not
gas—and are electrostatically charged, which causes them to stick
readily to other particles such as household dust, moisture droplets,
and smoke. They also cling to surfaces like walls, floors, and furniture.

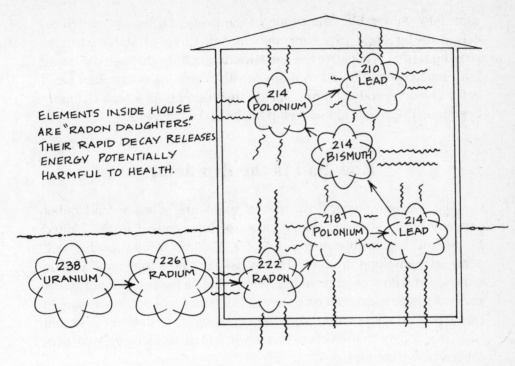

ELEMENTS INSIDE HOUSE ARE "RADON DAUGHTERS." THEIR RAPID DECAY RELEASES ENERGY POTENTIALLY HARMFUL TO HEALTH.

Illustration 15. Radon's "family tree"

Radon gas tends to be exhaled after being inhaled and so leaves little of its inherent radioactivity (which is slight) behind in the lungs. But the radon progeny that are always present with the gas, either free-floating or adhering to airborne particles, stick firmly to the lungs' mucous linings. There they bombard the tissue with powerful radioactivity primarily in the form of alpha particles, which are known to cause changes in human cells that medical researchers widely believe can lead to cancer. So strongly has the connection been established that the United States Environmental Protection Agency (EPA) classifies radon as a known human carcinogen.

If the bombardment were to strike external skin, the dead layers which are normally present would offer protection. Alpha particles are destructive because they are large (at least on an atomic scale), but because of their size they cannot penetrate deeply like other forms of radioactivity such as X rays. Even a sheet of cellophane is sufficient to deflect alpha particles.

But lung tissue contains no dead skin barrier, and in addition is especially vulnerable. The tissue of the bronchial linings, which receives the greatest exposure and the full force of alpha particles striking it (this is because the particles do not pass through the tissue but remain on and in it), contains mostly a type of cells called basal cells. These reproduce more quickly than others. As a result, tumors beginning there may also develop faster and more readily.

How Bad Is the Problem?

The Surgeon General warns that exposure to radon and radon progeny is the second leading cause of lung cancer in the United States and is possibly responsible for 5,000 to 20,000 deaths each year. Only smoking produces higher risk. But how long must you be exposed and how great must the radon level be for you actually to get cancer? There is no concrete answer. As with smoking, some people may develop cancer after only minimal exposure to low levels; others may live longer-than-average lives without ill effects despite prolonged heavy contact.

Radioactivity is measured in units called picocuries (pCi). Amounts are usually expressed as a number of picocuries per liter of air (pCi/L). Sometimes measurements are expressed in working levels (WL) instead; a level of 0.02 WL is usually equal to about 4 pCi/L in a typical home.

In the atmosphere, radon amounts average about 0.25 pCi/L. According to EPA calculations, a person exposed to this concentration of radon—or even up to as much as 2 pCi/L—for 75 percent of the time during a period of seventy years (this is what the EPA considers to be an average lifetime) has the same risk of dying from lung cancer as that of an average nonsmoker. Someone exposed to between 10 and 20 pCi/L for the same duration has a risk equaling that of a person who smokes one pack of cigarettes per day. Exposure to between 100 and 200 pCi/L is the equivalent of smoking four packs of cigarettes daily.

These figures are extrapolated from studies begun in the 1950s and 1960s and conducted for more than thirty years—in some cases, for more than forty years—on uranium miners and other miners in the United States, Canada, Australia, China, and Europe. Critics of the radon issue dispute whether comparing the conditions of exposure among miners to those of the general residential population is accurate, and it is true that a number of studies—perhaps a growing number, as some are still ongoing—have shown the need for increased caution in projecting results from one group to the other.

But radon levels even higher than in some uranium mines have been found in some homes, and most people in the United States spend more time inside their homes or in other buildings that can contain radon than do miners working one or sometimes even two eight-hour shifts per day, five days per week. In fact, a significant finding in the miners' studies, and one that corresponds to medical views of how cells can become cancerous, is that exposure to low amounts of radon over long periods of time produced greater lung cancer risk than high exposures over short periods.

To sum up, there is no doubt that radon and radon progeny cause lung cancer in humans and have caused it among underground miners in all significant studies. But it remains uncertain how to adjust the results of these tests so they accurately present the cancer risks faced by the general population. Tests are currently under way aimed at determining the risks from radon encountered in residential surroundings rather than occupational ones. Until their results are tallied, there simply are no others on which to base strategies for avoiding potential harm.

Testing

The average radon level in houses and other buildings is believed to be about 1.3 pCi/L. Currently, the EPA has established as a guideline an annual average of 4 pCi/L as the amount above which you should definitely act to reduce levels in your home. This figure does not represent a concentration of radon the EPA feels is safe. Rather, it represents a concentration that in most houses and other buildings is difficult and often prohibitively expensive to reduce by making structural modifications and by taking other radon-control steps. At an exposure to 4 pCi/L over a lifetime, the risk for nonsmokers of developing lung cancer from radon is about 2 in 1,000 (this is greater than the risk of dying in a house fire or of drowning). For smokers, the risk is about 29 in 1,000. The American Society of Heating, Refrigeration, and Air-Conditioning Engineers (ASHRAE), which also sets standards for indoor air quality, currently recommends 2 pCi/L as a guideline for radon concentration.

It is worth noting that, according to EPA figures, probably only 6 to 7 percent of homes in the United States have radon levels above 4 pCi/L, and only about 17 percent of homes have levels of 2 pCi/L and above. But the only way to find out how much radon your home contains is to test for it. Fortunately, this is easy and can be done in stages that involve little expense or time.

Every home should be tested. Afterward, decide whether you and your family can live comfortably with the risk shown by the results, and let that decision determine whether to attack the condition or leave it alone.

How to Test

There are two general categories of tests: short-term and long-term. Kits for both containing easy-to-use testing devices are available at hardware stores or by calling your state's radiation protection office or health department, an EPA regional office, or a local chapter of the American Lung Association. Costs, including the laboratory fee for evaluating data collected by the devices, range from about $10 for short-term testing to about $30 for long-term testing.

Sometimes discounts and even free testing devices are available from local governments and private organizations conducting special programs, so it is worthwhile to investigate such possibilities before making a purchase. But do not put off testing in order to wait for a bargain. When you buy a testing device, make sure it displays the phrase "Meets EPA Requirements." This means representative samples from the manufacturer have passed the EPA's Radon Measurement Proficiency Program (RMPP) tests.

To get the most accurate results for the least amount of money and effort, the EPA recommends doing a short-term test first; this takes between two and seven days and results are usually available within two weeks. If this test shows radon levels higher than 2 pCi/L but not exceeding 10 pCi/L, a long-term test is recommended, preferably one lasting a year. Because radon levels can vary from day to day and from season to season, long-term testing is more accurate than short-term testing.

If the radon level indicated by the initial test is higher than 10 pCi/L, you should do a second short-term test instead and average the results—waiting for findings from a long-term test may not be prudent. Use this method also if you need results quickly for other reasons; for instance, if you are planning to sell your home or are considering purchasing one and wish it tested beforehand.

If the results of an initial test show radon levels below 2 pCi/L, no further action is needed, but it might be wise to do a long-term test—or additional short-term tests once every three months for a year—to confirm the average level.

EVALUATING TEST RESULTS (Adapted from EPA)

Test Results (pCi/L):	Response:
Less than 2	Optional short-term or long-term testing to confirm results
2–4	Perform long-term test, mitigation action optional
4–10	Perform long-term test, take mitigation steps within two years
10–200	Perform two short-term tests and average results, take mitigation steps within months
Over 200	Perform two short-term tests and average results, take mitigation steps within weeks, consider relocating until radon levels are reduced, call EPA

If radon levels are between 2 pCi/L and 4 pCi/L, taking steps is probably not warranted. If levels are between 4 pCi/L and 10 pCi/L, you should take action within two years, sooner if levels are near the upper end of this range. If radon levels are between 10 pCi/L and 200 pCi/L, action should be taken within a few months. If levels are higher than 200 pCi/L, take action within weeks, and if this is not possible consider moving out of the building until levels can be reduced.

Call the EPA for special advice and possible assistance if your readings are this high.

Performing Tests Yourself

Testing methods and devices for amateurs are called passive because they require no skills to use. The most common short-term testing device is called a *charcoal monitor*. It consists simply of a small canister or pouch filled with activated carbon that absorbs radon. The monitor is placed in an area and left for a period of two to seven days, depending on the manufacturer's instructions. Then it is returned by mail for laboratory analysis and the results are mailed back within a few weeks, often within five days.

Because of their low price—about $10—it makes sense to use more than one monitor when conducting this type of test. Some manufacturers offer discounts when you order several monitors

Illustration 16. Charcoal monitor

at a time as a way of inviting this practice, but you should insure against the possibility of a faulty monitor or of a laboratory error by using monitors from different manufacturers. When you get the reports from each laboratory, average them yourself to obtain a final score.

In all cases, accuracy depends on following the manufacturer's directions càrefully. Plan to conduct the test in the lowest lived-in area of the house: the basement if it is frequently used, otherwise the ground floor. Keep windows and doors to this area closed for at least twelve hours before beginning the test. If you have appliances or plumbing fixtures in the area to be tested that use water from a well, shut off their water supply and do not use them for twelve hours prior to or during the test. This will eliminate water-borne radon as a possible source of high readings.

Do not unseal monitors before you are ready to begin testing. During the test, keep windows and doors to the testing area closed as much as possible and avoid using fans, range hoods, air conditioners, and other ventilation appliances located there (in addition to any appliances or fixtures that use well water, as was mentioned above). Do not conduct the test during stormy weather or a period of high winds. The overall strategy is to create a realistic worst-case scenario—the greatest possible accumulation of radon under ordinary conditions, not short-lived extremes.

If you're using two monitors, place them a few feet apart. The lab results from both should be virtually the same. If you're using a third monitor, place it on the next higher floor; expect its reading to be about 60 percent lower than the readings of the others. If your water comes from a well, you can conduct a preliminary test for water-borne radon by placing the third monitor in an upstairs bathroom, away from the main test area. High readings there indicate the need for further testing of the water (which is described farther on in this chapter).

Select an area for the monitors that is used regularly, but except when you're making a preliminary water test with a third monitor do not place them in a kitchen or bathroom, where temperatures and humidity levels tend to be higher than in other rooms.

Set monitors at least twenty inches above the floor and eight inches below the ceiling, in places where they will be exposed but not disturbed; good places are on an open shelf or hanging from an exposed ceiling joist. Do not place monitors behind large items of furniture or sandwich them between objects like books that are closely spaced and may block the flow of air. Keep the monitors from resting against an exterior wall; also avoid placing them on a mantel or in any location where they will be susceptible to drafts, high heat, or high humidity.

Leave each monitor in place for as long as the manufacturer's directions indicate. Then reseal it, enclose it in the package or mailing envelope supplied with it, and return it immediately to the address specified.

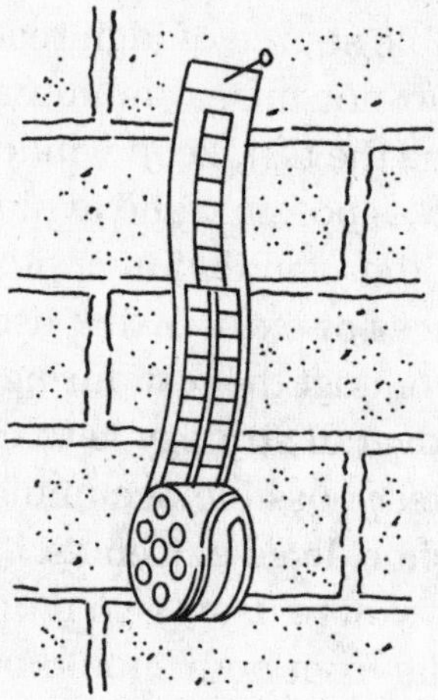

Illustration 16A. Alpha-track detector

For long-term testing the most common device is an alpha-track detector. It consists of a small container with a piece of plastic inside that is sensitive to marking by the alpha particles released by radon and radon progeny.

Using an alpha-track detector is actually easier than using a charcoal monitor. Except when a test lasts three months or less, normal living conditions are called for—including ordinary use of windows, doors, and ventilation items; and detectors are not affected by elevated temperatures and humidity. You must keep the detector away from strong drafts, and the location must insure that the device remains undisturbed for the duration of the test, but those are the only real demands. If you are planning major remodeling work or are replacing heating or central air-conditioning equipment, delay long-term radon testing until these jobs are completed.

Alpha-track detectors typically cost about $25. As with charcoal monitors, and also because of the investment in time that long-term testing involves, it is wise to buy and use more than one detector at a time. Place them as described for charcoal monitors. If your household includes children under twelve years of age, it is especially important to measure long-term radon concentrations in their bedrooms and in other areas where they spend a lot of time. Evidence suggests that children may be more sensitive than adults to the effects of inhaled radon decay products.

It is also possible to buy continuous radon-tracking devices that are electronically powered and show results either on a liquid crystal display (like that of an electric clock) or on a roll of paper (like a cash register receipt). As you might expect, these testers are expensive; but while they might seem to have apparent advantages over spot testers like charcoal monitors and alpha-track detectors, actually the continuous availability of data they provide does not by itself show a true picture of a home's radon concentration. That data, too, must be gathered and averaged just as with charcoal and alpha-track testers.

Having Tests Performed by a Professional

Professional radon testing is available if you do not want to do the tests yourself, or cannot; for example, if you are buying a house and need or would like it tested beforehand. Real estate transactions generally call for radon testing to be done only by professionals with qualifications accepted by the state or other authorities. To find a qualified professional testing company, look in the telephone book or contact your state radon office or regional EPA office for a list of companies in your area that have passed the RMPP tests (mentioned earlier). Companies and individuals listed in the RMPP must follow EPA measurement procedures and demonstrate their ability to take accurate measurements with specific testing devices. Be sure the individual tester, not merely the head of the testing company, has the qualifications mentioned above. Some states have additional requirements for professional radon testers.

Before deciding on a testing company, compare information from two or three others—as you would do when selecting any contractor. Besides asking about rates and qualifications, find out what the tests entail. Many consist of nothing more than having a technician place and later remove passive devices like charcoal monitors and alpha-track detectors as described above. But professionals can also perform so-called active tests that require special skills and sophisticated equipment. These can identify radon levels immediately and can pinpoint the places where radon enters a home.

Getting Rid of Radon

If testing your home reveals elevated radon levels, the EPA recommends that you have a qualified radon-mitigation contractor fix the problem. This is because lowering high radon levels usually requires specific technical knowledge and special construction skills. If the job is not done correctly, the results can be worthless or even increase radon levels.

As in finding a qualified radon tester, the best way to find a qualified radon-mitigation contractor is to call your state radon office or nearest EPA regional office. They should be able to provide a list of contractors in your area who meet the requirements of the EPA's Radon Contractor Proficiency Program (RCPP), a series of training courses and examinations. Like companies and individuals listed in the RMPP, described above, contractors listed in the RCPP must follow EPA guidelines for minimum quality standards. Currently, you can expect to pay an average of $1,000 to $1,500 for routine professional mitigation, according to the EPA.

Basically, two strategies are available for ridding a home of radon: preventing the gas from entering, and reducing the concentration of gas that does enter. Preventing entry is the most widely used strategy because of its high success rate and frequent cost-effectiveness, and because it can be implemented in a majority of homes. Most often, a technique called subslab depressurization is used. By this method, radon-laden air is drawn directly from the soil beneath a house by means of a suction fan and is exhausted safely to the outside through a pipe. The process creates an area of low air pressure around the foundation. As indoor air under greater pressure migrates toward this zone, the outflow of air from the basement into the soil seals most foundation and crawl space openings, eliminating the need for extensive sealing with caulking compound and other materials.

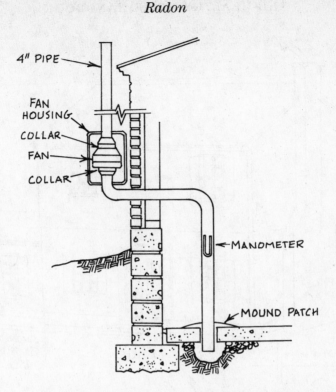

4" PIPE

FAN HOUSING

COLLAR

FAN

COLLAR

←MANOMETER

←MOUND PATCH

Illustration 17. Sub-slab depressurization system

Methods that reduce radon concentrations after entry usually cost the same or more than typical subslab systems and generally are effective only when radon levels are fairly low (15pCi/L or less) to begin with. Of course, natural ventilation is free and sometimes works, but this method is impractical in most parts of the country, even where weather is warm year-round. The most common post-entry method of radon reduction is installation of a heat-recovery ventilator, or HRV, a type of blower that can exhaust radon-laden air from a basement or crawl space to the outside without wasting indoor heat that might otherwise escape with it. Selecting and installing a heat-recovery ventilator for radon reduction is a job for a licensed heating contractor working under the supervision of a certified radon-mitigation contractor. (Heat-recovery ventilators are fully described in the chapter "Improving Ventilation.")

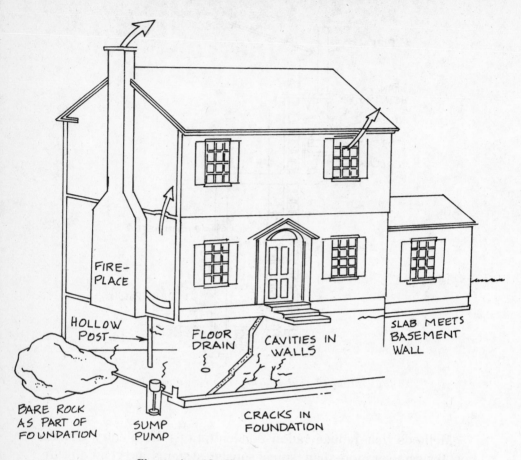

Illustration 18. How radon can enter a house

Sealing Radon Entry Routes

While meticulous sealing of basement or crawl space openings through which radon can enter is not necessary for the success of either subslab depressurization or ventilation methods, closing as many entry points as practical makes sense. Professional radon mitigators generally agree that the best way to proceed is to seal the largest and most accessible entry points first, then install or operate whatever system you decide on and test the results. Even if subsequent testing still shows high radon levels, it is likely—at least with subslab depressurization systems—that desired results can be

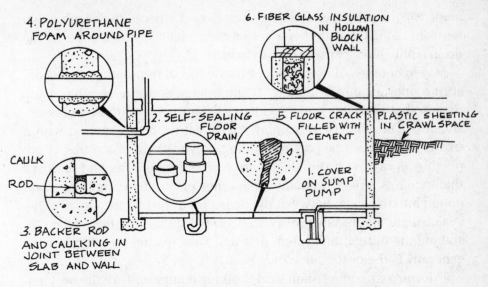

4. POLYURETHANE FOAM AROUND PIPE

6. FIBER GLASS INSULATION IN HOLLOW BLOCK WALL

2. SELF-SEALING FLOOR DRAIN

5. FLOOR CRACK FILLED WITH CEMENT

PLASTIC SHEETING IN CRAWL SPACE

CAULK

ROD

1. COVER ON SUMP PUMP

3. BACKER ROD AND CAULKING IN JOINT BETWEEN SLAB AND WALL

Illustration 19. Methods for sealing radon entry points

achieved more easily and quickly by modifying the system than by attempting to block more openings.

A logical place to begin is by closing the sump opening, if there is one. Use a plastic sump cover that can be sealed over the opening with adhesive caulking compound (use silicone adhesive; it is easier than other kinds to remove later if necessary), or else buy a plastic sump collar or basin with a removable top. The collar or basin fits into the sump opening and is sealed around the perimeter with foam or caulk. The top bolts to the rim.

If the sump contains a pump, buy a cover that has grommeted openings for the discharge pipe and the pump's electric cord. If the pump extends above the sump opening, replace it with a low-profile submersible pump that fits beneath the cover. Sump covers with grommeted openings are also available with an additional hole for a radon-venting pipe installed as part of a sump hole depressurization system, described later.

Another opening to close is the floor drain. If the drain is no longer used, one solution is simply to cover it with a piece of galvanized sheet metal cemented to the floor with adhesive caulk. But a better fix is to replace the existing drain with a self-sealing floor drain made for the purpose. These remain closed, preventing radon-laden air from rising through them, unless the weight of water entering the drain from above causes them to open.

Open courses of hollow block at the top of foundation walls constitute another large opening. If the openings are accessible, pack them to between one and two inches from the top with fiberglass insulation. Then fill the remaining cavity with mortar, applied with a trowel, or with a one-part expanding polyurethane foam that is applied by spraying. If the blocks are covered by boards laid flat over the openings, inject urethane or another elastomeric caulking compound into the seam between the boards and the blocks. Usually, this is a difficult and tedious procedure, one of those situations when installing the mitigation system first and then testing its effectiveness generally is the best course of action.

Mortar, expanding foam, and caulking compound are the best materials for sealing openings in walls, including open courses of hollow block at the top of foundation walls, and openings around plumbing pipes, electrical conduits, window and door frames, and appliance vents. (Although some of these may be above ground, hollow blocks next to them can conduct radon from below.) Mortar is best for filling large openings. To avoid cracking, use mortar in the form of hydraulic cement, latex-concrete patching compound, or a patching compound containing fibers that reduce shrinkage. Polyurethane foam is more expensive and produces a less-attractive surface, but it is often easier to apply and it creates a more durable seal. When applying polyurethane foam in large quantities, wear a protective mask to avoid breathing fumes and ventilate the basement adequately to keep fumes from entering the house.

Use caulking compound for filling narrow cracks. Urethane and other high-quality elastomeric caulks are recommended. Before applying caulk, check the directions on the tube for the minimum crack width recommended for ideal bonding. In most cases, cracks less than a quarter-inch wide require chiseling or grinding with an abrasive disk grinder to widen them sufficiently (EPA guidelines suggest enlarging cracks to half an inch wide and half an inch deep). Cracks less than an eighth of an inch wide and shorter than eight feet can be left unsealed.

Radon in Water

Water-borne radon is a problem only when water is supplied directly from wells or underground springs. Water from reservoirs, lakes, rivers, and most municipal sources is virtually free of radon because it is exposed to air either at its source or during treatment.

If tests of the air in your home reveal high levels of radon, and if your water comes from underground, you should have the water tested. The EPA estimates that the average radon level in groundwater is about 800 pCi/L; however, ingesting water containing radon is not considered a health risk because a level of 10,000 pCi/L in water is required to liberate 1 pCi/L of radon or radon progeny into the air.

For the best results, send a sample of your water to a laboratory. To find one, call your state radon office, your local health department, or a regional EPA office. The laboratory will send you containers and instructions for taking samples. Generally, the procedure involves filling the containers with water in a way that loses as little radon as possible to the air. The water should be allowed to run for about ten minutes before sampling, the faucet must not be equipped with an aerator, the container must be filled slowly to avoid getting air bubbles into the sample, and the container should be closed immediately when full. Mail the containers back to the lab. The results should come back within a few weeks. Do-it-yourself kits for testing the radon content of water are available, but their results may not be accurate because of the negative effect of humidity on the testing materials.

The standard set by the federal government in 1993 for radon in water is 300 pCi/L. While reduction systems are available to reduce radon in water to far below this level, you need to weigh their expense (anywhere from $1,000 to around $6,000, according to recent EPA surveys) against the relatively small hazard water-borne radon presents. Generally, it makes sense to consider reducing water-borne radon only when levels approach or exceed 10,000 pCi/L.

Two methods are commonly used to reduce radon in water: filtration and aeration. The former removes radon by absorbing it in a filter medium, normally granulated carbon (charcoal). The latter frees radon from water into air that is then directed safely to the outdoors.

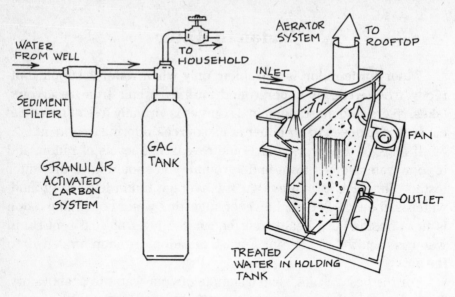

Illustration 20. Filter and aeration systems

Both systems are best installed by professionals under the supervision of a certified mitigation contractor.

Filtration is the most practical method for residential application where radon levels in the water are relatively low (up to 10,000 pCi/L). A filtration system consists primarily of a prefilter—for removing sediments from the water—and a fiberglass tank, similar to tanks used for water softening, filled with one to three cubic feet of granular activated carbon (GAC). For best results, the tank should hold approximately fifteen gallons of water on top of the carbon layer.

The prefilter and GAC tank are connected to the incoming water supply at a convenient location near the water's point of entry into the house but beyond the pressure tank. Water passes from the pressure tank through the filter and then enters the GAC tank at the top. It is drawn off at the bottom.

Theoretically, certain precautions are necessary with a GAC filtration system. Because radon and radon decay products accumulate in the tank, there is a potential need for shielding the tank to protect the home's occupants from gamma radiation, which can pass through the tank and into the air. (Alpha and beta radiation, the other radioactive components of radon, remain trapped inside the tank, where they quickly decay.) Such shielding, while seldom needed, can be accomplished by enclosing the tank in a compartment built of concrete blocks. Also, where climate permits, the tank can be located in a tank house built away from the residence. Gamma radiation escaping from a tank becomes negligible at a distance of four to ten feet.

Another potential problem concerns disposing of the carbon when it can no longer absorb impurities from the water or when it develops high levels of radioactivity. Normally, replacing the prefilter cartridge at intervals specified by the manufacturer prevents the GAC system from clogging by most sediments, but the carbon can harbor bacteria and other organisms that may inhibit filtration and contaminate water. Replacing the carbon annually, or at intervals of up to five years if contamination does not occur, sanitizes the tank and at the same time assures that the carbon does not become so radioactive that special disposal techniques are required.

For radon concentrations above 10,000 pCi/L, an aeration system generally is recommended. Such a system consists of a tank into which incoming water is either sprayed from a nozzle to mix it with air or else is allowed to enter and is then saturated with air directed upward from below. The aerated water loses its original pressure; this is restored by means of a pump. The water then is available for use—or, to reduce extremely high radon levels, it can be directed to a GAC unit.

Some aeration systems use air from the system's location (usually the basement) to supply the tank. Other systems employ fresh air from outdoors. In all cases, radon-laden air leaving the aeration tank is ducted outdoors through a pipe extending to the roof level of the house, just as with a subslab depressurization system. Because of its many parts, some of which have small passages prone to clogging, an aeration system requires professional cleaning and maintenance at least yearly.

Combustion Products

What Are They?

IN MANY HOMES—PROBABLY IN MOST—GASES AND OTHER particles created as the result of burning, and particles that are or were living organisms, contribute more significantly than any other substances to indoor pollution. These contaminants are not found in building materials. Instead, they are by-products of living in a house or are unavoidably present in the air.

Most combustion-related products are emitted from improperly installed or poorly functioning fuel-burning appliances and fireplaces. Examples are furnaces, kerosene heaters, and gas-fired kitchen ranges, water heaters, and clothes dryers. Tobacco smoking by a house's occupants can also generate significant amounts of these products.

Carbon Monoxide

The most dangerous combustion product, and one of the most common, is carbon monoxide (CO). This colorless, odorless gas is the leading cause of death by poisoning in the United States, according to separate research published in both American and Canadian medical journals. Produced by incomplete burning of fuel, CO is a chemical asphyxiant that bars oxygen from the bloodstream, thereby starving brain cells and others—eventually killing them—and interfering with respiratory, metabolic, and other processes that sustain life.

The ways of carbon monoxide are insidious. Normally, oxygen attaches to hemoglobin in the blood and is carried throughout the body. However, carbon monoxide attaches to hemoglobin about two hundred times more readily; therefore, even when both oxygen and carbon monoxide are present, the blood will pick up and transport CO throughout the body in preference to oxygen.

Once in the bloodstream, CO forms a substance called carboxy-hemoglobin (COHb). Initial symptoms of CO poisoning, usually a mild headache, generally appear in adults when the level of COHb in the blood reaches about 15 percent (cigarette smoking typically produces levels around 9 percent). Nausea and severe headache occur when COHb levels reach about 25 percent.

Victims of CO poisoning at COHb levels below 30 percent usually remain conscious and recover fully in five to eight hours after breathing fresh air. But at COHb levels higher than 30 percent, long-term ill effects are probable, including brain damage and harm to the fetus if the victim is pregnant. COHb levels of around 45 percent produce unconsciousness; death results if exposure continues or if COHb levels rise toward 50 percent.

How quickly CO enters the blood depends largely on the concentration of the gas in the air. At a concentration of 100 parts per million (ppm), the time required for the blood of an average adult to develop COHb levels of 10 percent is about ninety minutes. At a concentration of 200 ppm, the time required to reach the same percentage is only thirty-five minutes; in air containing 400 ppm CO, the percentage of COHb will rise to 10 percent in about fifteen minutes.

Carbon Dioxide

Carbon dioxide (CO_2), which is exhaled from the human body by breathing and is familiar to most people as the source of bubbles in soft drinks, is also a combustion product. Like carbon monoxide, carbon dioxide is an asphyxiant, but it also is a respiratory stimulant that causes breathing to become faster. Where carbon dioxide is produced as a result of combustion, other compounds—particularly CO—usually are present and supercede it in causing poisoning symptoms. However, the effect of CO_2 on breathing can cause victims to inhale more toxic compounds and at a faster rate.

Nitric Oxide and Nitrogen Dioxide

Nitric oxide (NO) and nitrogen dioxide (NO_2) are pollutants produced by the burning of fossil fuels. Nitric oxide is colorless, odorless, and tasteless. Like CO and CO_2, it readily attaches to hemoglobin and thereby displaces oxygen from the bloodstream. Nitrogen dioxide is a corrosive, suffocating gas that is painful to inhale. Its chief effects are to cause lung damage and to increase susceptibility to bacterial lung infections. People with asthma can be affected by very low levels of NO_2, and children who are chronically exposed to even low levels appear to have higher rates of respiratory illness than other children who are not exposed. Adults chronically exposed to moderate or higher levels show substantial pulmonary changes and may develop emphysema. Unfortunately, although many effects on cells of NO_2 exposure are immediate and can be easily documented, larger biological effects may be delayed. So far, this has complicated complete understanding of the effects of these compounds.

Sulfur Dioxide

Sulfur dioxide (SO_2) is also a combustion product with a sharp, distinctive odor—like that of rotten eggs. Because it is very soluble in water, it quickly exerts its effects after entering the body, typically by irritating the eyes, nose, throat, and other areas of the upper respiratory tract. (On the other hand, NO_2 is not easily dissolved in water; consequently, it travels farther into the body—to the lower respiratory tract—before delivering effects.)

Asthmatics typically experience discomfort from sulfur dioxide at lower concentrations than people with unimpaired breathing functions. However, at moderate to high concentrations even healthy people can develop asthma symptoms—wheezing and chest tightness due to narrowing of the lung passages.

Particulates

Particulates—or, simply, particles—given off by combustion sources are of special concern. They can have effects of their own or can act as vehicles to which other combustion products, and particles such as radon, become attached. The danger from particles depends largely on their size. So-called large particles, greater than ten microns in diameter (a micron equals a thousandth of a millimeter), typically are trapped in the nose and eventually are expelled. The same is true for about 60 to 80 percent of mid-size particles, ranging in diameter from five to ten microns. The remainder pass into the lungs, as do virtually all particles less than five microns in diameter.

Protective fluids in the lungs usually transport or dissolve the smallest of these, but those that are small enough to be inhaled yet are too large for protective fluids to eliminate can remain to form deposits that in turn can produce adverse health effects.

Such particles are called respirable suspended particulates (RSPs). There are hundreds of RSPs, many virtually unstudied. Among the most common and dangerous are polynuclear or polycyclic aromatic hydrocarbons (PAHs); these are carcinogenic. Others typically fall into categories of toxic trace metals, nitrates, and sulfates. Lead dust, asbestos, and radon are also RSPs.

Water

Water (H_2O), too, is a combustion product. While not directly harmful, water vapor released indoors can condense on cold surfaces such as windows and dampen surfaces and insulation inside walls. This in turn can cause structural damage and provide a breeding ground for molds, bacteria, fungi, dust mites, and other biological contaminants, all of which can have negative health effects, especially in allergic individuals.

How Bad Is the Problem?

Carbon monoxide from household combustion appliances causes nearly 300 deaths annually, according to the Consumer Product Safety Commission (CPSC). This figure can be considered a minimum estimate because the CPSC receives for evaluation only death certificates reporting fatalities believed by coroners and medical examiners to be product related. The Mayo Clinic, by contrast, has published reports that accidental exposure to carbon monoxide in the home is responsible for some 10,000 medical visits and 1,500 fatalities annually. Research at a Louisville, Kentucky, hospital during the 1980s revealed that nearly 25 percent of patients with flulike symptoms actually were victims of carbon monoxide poisoning; it also is believed that 70 percent or more of fire-related deaths in residences are due to CO inhalation from smoke rather than to burns (this figure includes victims who are incapacitated or rendered unconscious by the gas and therefore cannot escape the flames). On average, according to CPSC figures, deaths in the United States from smoke inhalation average between 4,000 and 5,000 annually.

The effects of other combustion products are less extreme, or at least less well-known. As was mentioned earlier, people with asthma may develop symptoms at far lower concentrations than do others. Figures published by the EPA show that asthmatics are responsive to nitrogen dioxide in the air at concentrations of about 0.5 parts per million (ppm), while normal, healthy adults may not respond until concentrations are at or above 2 ppm; however, at high enough concentrations or over long periods of time, symptoms and eventual lung damage are virtually inevitable for everyone.

Testing

Practically every home needs a carbon monoxide detector for safety. Because the gas cannot be seen, smelled, or tasted, a detector is the only means of testing for it; not to have one can have fatal consequences. CO detectors are essential in homes with fuel-fired furnaces, fireplaces, woodstoves, and/or unvented kerosene or other fuel-burning heaters. They are absolutely vital (and are often required by code) in new, tightly built, heavily weatherstripped homes with any of these devices; such construction has a tendency to starve combustion appliances of air, causing them to burn incompletely and produce more carbon monoxide and other pollutants than normal. There is also a higher risk that backdrafting can occur, which draws exhausts from combustion appliances back down their flues or chimneys. (Backdrafting is described farther on in this chapter.)

But even if your home contains no combustion heating appliances, carbon monoxide can be produced by a malfunctioning gas kitchen range, by automobile exhaust in an attached garage, and

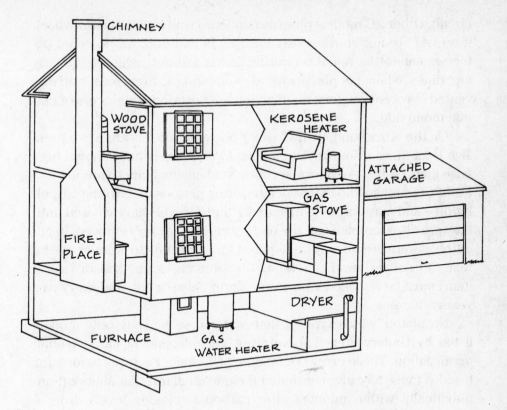

CHIMNEY

WOOD STOVE

KEROSENE HEATER

ATTACHED GARAGE

GAS STOVE

FIRE-PLACE

FURNACE

GAS WATER HEATER

DRYER

Illustration 21. Common carbon monoxide sources

from combustion devices on the other sides of apartment walls in multi-family buildings.

Carbon monoxide detectors resemble smoke detectors in appearance and function. At this writing, they range in price from $40 to $80. A detector sounds an alarm when CO levels are dangerously high for a short period of time (usually 400 ppm CO for fifteen minutes) and when low concentrations persist for a correspondingly longer amount of time (100 ppm CO for ninety minutes). The noise they make usually is louder and more piercing than that of a smoke detector so it will rouse victims from the stupor that normally accompanies CO poisoning.

Battery-powered and AC-powered CO detectors (these operate on household electricity) are available, as are AC detectors with battery backups. All have advantages and disadvantages: the latter are recommended but generally must be wired directly to a household circuit. Other AC models plug into ordinary wall outlets—a drawback if outlets are not conveniently located; in addition, AC-powered detectors cannot be relied on during power failures, which frequently are times when people resort to woodstoves, fireplaces, and unvented kerosene heaters, all of which are notorious producers of carbon monoxide.

On the other hand, while battery-powered models are less expensive than most AC-powered models, older models have shown high false alarm rates—or, more accurately, oversensitivity to minute levels of CO such as those that occur during periods of extreme air pollution—and may continue to sound after CO levels have cleared until they are disassembled and the battery and sensor are removed. In addition, several hours may be required for such units to recover and be ready to deliver another reading; in some cases, the sensor (a $20 item) must be replaced after every alarm, false or not, and every two years regardless.

No matter which type of detector you select, buy only models listed by Underwriters Laboratories, an independent safety-testing organization. These carry the familiar "UL" logo. Features to look for besides those already mentioned include an alarm that shuts off automatically within minutes after carbon monoxide levels drop; a manual "hush" or reset button that can be pushed to silence the alarm briefly in the presence of CO (useful for detectors located near a kitchen stove or near a fireplace that may generate high CO levels only during lighting); a digital display or warning light; a light indicating that power is on; and a test button like that on a smoke detector, to verify that the detector is working. A new type of AC-powered CO detector contains a microchip processor that actually counts CO molecules landing on the sensor. When levels of CO exceed 100 parts per billion within a time period of 90 minutes an alarm sounds and a red light on the device glows. At CO levels of 50 parts per billion within a 45-minute period, the detector emits a yellow warning light. The processor resets automatically after a detection event.

For basic protection, install one or more CO detectors where they can be heard from all bedrooms with the doors closed. For more thorough protection, also install detectors near potential sources of CO—the furnace room, garage, fireplace, and any gas appliance such as a kitchen range or clothes dryer. (Do not install CO detectors in rooms

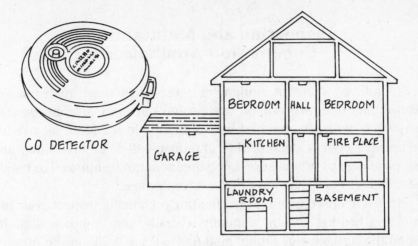

Illustration 22. Locations for carbon monoxide detectors

containing these items; momentarily high CO levels might sound an alarm.) Also consider installing detectors in rooms where household occupants spend large amounts of time—living rooms, family rooms, and dining rooms, for example. Clean, test, and maintain detectors frequently, according to the manufacturer's instructions.

Without a CO detector, the only way to tell whether the gas is present is to watch for symptoms of CO poisoning. Besides the key symptoms described earlier, other symptoms of carbon monoxide poisoning at low levels mimic flu and mild food poisoning—malaise, chills, slight fever, unexplained diarrhea or nausea. At moderate levels, symptoms include disorientation, dizziness or grogginess, and a reddish complexion. The key is that symptoms disappear about four hours after exposure ceases, usually because the victim has left the premises. Certainly, if you or other household occupants suffer symptoms at home that go away at work or school, a doctor should be consulted and a blood test for carboxyhemoglobin level should be considered.

Testing for combustion products other than CO and for biological contaminants is impractical. Fortunately, health effects from these pollutants are seldom life-threatening and, except for biological contamination, usually are signaled first by increased carbon monoxide levels, which can trigger an alarm.

Inspecting and Maintaining
Combustion Appliances

Overall, combustion equipment must receive adequate air for burning fuel and be installed and sealed correctly so that exhaust cannot mix with air breathed by occupants. In addition, air needed for burning cannot deplete the amount needed for breathing, and heat produced by combustion appliances cannot be allowed to create either indoor condensation or excessive dryness.

Have a licensed professional heating technician inspect your furnace and heating system annually. Usually, the company that installed the furnace can supply qualified service; look on the furnace for the name of the installer. If the company has gone out of business or if you decide to seek service elsewhere, choose a heating technician as you would any contractor: question three or more prospects. Ask for proof of liability and workers' compensation insurance, find out how long they have been in business and what professional organizations they belong to, and get names of references you can call. Licensing of technicians varies among states, and there is no universal industry qualification; however, you might consider calling your local consumer affairs department to find out what qualifications are needed for heating technicians in your community.

Oil Furnaces

For an oil furnace, inspection and routine maintenance should include checking for air leaks; cleaning and servicing the filter system, combustion chamber, combustion air blower, burner motor, and firing system (burner nozzle and ignition electrodes); and also adjusting the draft regulator on the stack (the exhaust pipe leading from the furnace to the chimney), cleaning the chimney base, inspecting all safety devices—these usually include the burner-disconnect switch, primary control (also called the stack control) and reset switch, both mounted on the stack—and checking the fuel storage tank(s) and supply line(s).

The technician should inspect the burner flame for size and color during operation. This determines how efficiently the furnace operates. For greater accuracy you can request that additional flame

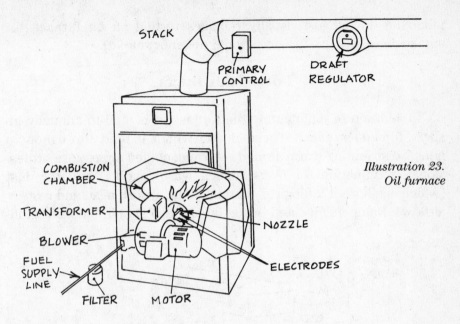

STACK

PRIMARY
CONTROL

DRAFT
REGULATOR

COMBUSTION
CHAMBER

TRANSFORMER

BLOWER

FUEL
SUPPLY
LINE

FILTER   MOTOR

NOZZLE

ELECTRODES

Illustration 23.
Oil furnace

testing be done with instruments that gauge smoke density, carbon dioxide content, and exhaust temperature. To check for cracks in the heat exchanger—these can mix potentially deadly combustion gases with indoor air and are especially dangerous in a forced-air system—order a tracer gas test. If cracks are found, replace the exchanger or the furnace immediately (see "Healthy Heating Systems" beginning on page 102).

Gas Furnaces

Inspecting and maintaining a gas furnace are less rigorous than the same chores for an oil furnace but should include annual tracer gas testing to check for heat exchanger cracks. Also included should be vacuuming the inside of the furnace, cleaning and adjusting the burners, and checking the pilot light or ignition system, burner flame controls, thermocouple, gas pressure regulator (also called the combination switch), and plumbing (including valves) supplying the furnace. It is especially important to check the draft hood and the chimney base to make sure both are unblocked. As with an oil furnace, testing by eye and with instruments can determine combustion efficiency. With a gas furnace, instrument testing to measure carbon monoxide in exhaust usually replaces testing an oil furnace for smoke density (a gas furnace should not emit smoke).

Forced-Air Ducts

In addition to maintaining the furnace, it is vital to annually inspect forced-air ducts. These ducts, which extend throughout a house, distributing warm air and returning cooled air, require professional examination for leaks (which waste energy) and for dust (which can harbor biological contaminants, absorb heat, and restrict airflow). Dampers (located inside ducts) also should be profession-

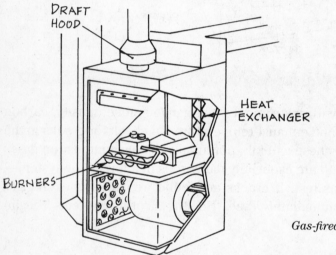

Illustration 24.
Gas-fired forced air furnace

ally adjusted, or balanced, to ensure uniform heating and cooling, and to avoid air-pressure imbalances that reduce overall efficiency.

Duct systems typically have either a filter or an electronic cleaning device located where returning air is recirculated past the heat exchanger. Filters should be cleaned or replaced every four to six weeks when the system is in use (this means year-round if the ducts also supply air from a central air conditioner or heat pump); air cleaners should be serviced according to the manufacturer's recommendations. (Electronic air cleaners are described farther on in this chapter.)

Filters and air cleaners for ducts are designed for easy replacing

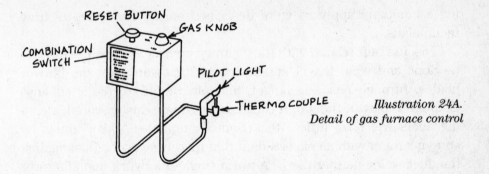

RESET BUTTON
GAS KNOB
COMBINATION SWITCH
PILOT LIGHT
THERMO COUPLE

Illustration 24A.
Detail of gas furnace control

or servicing by homeowners. When replacing a filter, use the same kind as the original to avoid altering the performance of the entire system, or else consult a heating technician for advice on other acceptable filters. Like regularly changing the oil in an automobile, frequently changing the air filter in a forced-air heating and/or cooling system is the single most effective step a homeowner can take to maintain the system's overall condition and the quality of the air it handles.

Have ducts cleaned every three to five years. Duct-cleaning companies are listed in the telephone book; the same caveats for choos-

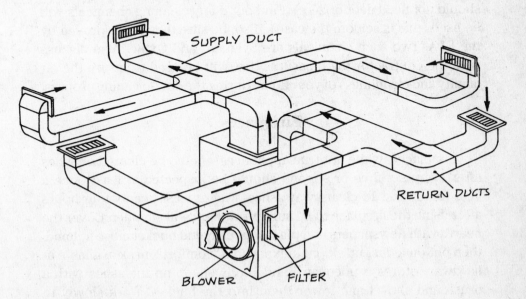

SUPPLY DUCT
RETURN DUCTS
BLOWER   FILTER

Illustration 25. Blowers and ducts

ing a company apply as were described earlier for hiring heating technicians.

Discuss beforehand with the cleaning contractor how the job will be done and what it will entail. Specify that you want the blower blades, furnace heat exchanger, and air-conditioner coils (if any) cleaned as part of the job. If these parts remain dust-covered, cleaning ducts will have little effect. Some companies "clean" ducts by spraying them with an encapsulant that merely seals the dust against the duct walls; neither the EPA nor a single ductwork manufacturer subscribes to this method. Instead, choose a company that employs one of two removal methods: scrubbing by hand with long, flexible brushes; or blasting with "skipper balls" sprayed into the ducts by compressed air. Both methods scour dust from the duct surfaces and collect it by means of a powerful vacuum cleaner attached at the ducts' lowest point.

The cost of cleaning ducts and furnace parts in a three- or four-bedroom house currently averages about $300, but it can go higher depending on the complexity of the ducts' design and the number of vents in rooms. Beware of add-on charges; some contractors may give you a basic fee and then attempt to attach extras such as disinfectants and deodorizers, which you don't need (disinfectants may not penetrate seams adequately to be effective, and properly cleaned ducts should not need deodorizers). Find out if any cleaning chemicals will be used (this is seldom the case); if so, be sure they are approved by the EPA. Two such chemicals are Sylguard and Oxine. Also discuss with the contractor the measures that will be used to insure the air quality and cleanliness of your home during the duct-cleaning process.

Fireplaces

Fireplaces, while not true appliances, should be cleared of ashes after every six fires or so, and should be inspected at least once a year for cracks. To clear away ashes, wait at least twenty-four hours after using the fireplace to be sure no embers still smolder. Cover the hearth with newspapers and place a large metal bucket close at hand; then put on a dust mask (an inexpensive "comfort" mask available at hardware stores is adequate) and gently scoop up the ashes with a brush and shovel and lower them into the bucket. Work slowly to avoid scattering the ashes and raising puffs of soot. Vacuum a fireplace only with a heavy-duty vacuum cleaner or with a vacuum designed for the purpose. Never use an ordinary household vacuum; soot can clog the motor and be exhausted by the machine throughout the house. Store removed ashes outside in a metal garbage can with a tight-fitting lid to be absolutely sure they are cool. Dispose of them either by adding them to your regular garbage or by spreading them as garden fertilizer.

Some fireplaces, especially those in older houses, have an ash dump or pit beneath them into which ashes can be raked through an opening at the back of the hearth. A clean-out door is located at the base of such a pit, either in the basement or outside at the bottom of the chimney. Empty the ash dump regularly (once or twice a year may be sufficient) and check that the door closes tightly.

To locate cracks, you may have to clean the fireplace thoroughly by scrubbing the inside with a bristle brush. This can produce clouds of soot, so wear a hat, gloves, tight-fitting goggles, and long-sleeve clothing. Eliminate the possibility of drafts by closing the chimney damper and any windows and doors in the room containing the fireplace. Also make sure the area surrounding the fireplace is protected against dust.

Any cracks wide enough to accept a knife blade require attention. You can repair cracks yourself without too much trouble, but if damage is extensive, hire a qualified fireplace mason to examine the situation; the fireplace may need renovating, which usually is a job for a professional.

Repairing Fireplace Cracks   To repair fireplace cracks, widen them to at least an eighth of an inch with a sharp cold chisel and either a hand-drilling hammer or a heavy ballpeen hammer. Do not use an ordinary claw hammer when striking a chisel; the hammer can splinter, sending metal shards in your direction. Wear eye protection in any case. When chiseling inside the fireplace, work carefully to avoid cracking firebricks, which are brittle.

Brush and vacuum the cracks free of chips and dust. Fill cracks between firebricks with refractory mortar; this comes in a cartridge and can be applied like caulking compound. Follow the manufacturer's directions. On the outside of the fireplace, use ordinary cement mortar mixed with water. Dampen the masonry first and then apply the mortar with a tuck-pointing trowel. Wear gloves when working with mortar; chemicals in the cement can irritate skin.

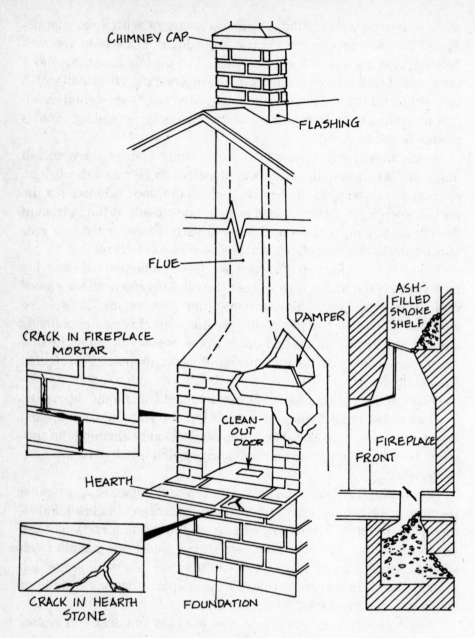

CHIMNEY CAP

FLASHING

FLUE

DAMPER

ASH-FILLED SMOKE SHELF

CRACK IN FIREPLACE MORTAR

CLEAN-OUT DOOR

FIREPLACE FRONT

HEARTH

CRACK IN HEARTH STONE

FOUNDATION

Illustration 26. Fireplace flaws

In both cases, press the mortar firmly into cracks and smooth it flush with the bricks. Tool the surfaces to resemble their neighbors when the mortar is stiff enough to retain a thumbprint when pressed.

Afterward, mist the fresh mortar regularly or tape plastic sheeting over it to keep it damp until it is completely cured, which takes about three days.

Cracked bricks, whether inside or outside the firebox, should be replaced. You can perform this task, too, but it is more difficult than repairing cracks. Begin by chipping out the damaged brick with a hammer and chisel, being careful not to break neighboring bricks. Buy a replacement brick at a building supply store or, if necessary, from a fireplace and woodstove dealer.

Vacuum the cavity where the new brick will go and dampen it and the brick with water. With a small mason's trowel, coat the back and sides of the cavity with a layer of appropriate mortar similar in thickness to the neighboring joints, and then spread mortar on the sides, ends, and back of the brick also. Press the brick into the cavity, aligning it with neighboring bricks and gently tapping the front surface with the end of the trowel handle to seat it correctly. Excess mortar should squeeze from the joints on all four sides; if it doesn't, remove the brick, apply additional mortar, and bed the brick again. When the brick is correctly placed, scrape away the excess mortar and gently clean the surfaces adjacent to the new joints with a damp sponge. Tool the joints and allow the mortar to cure as was described above.

Correct Use of a Fireplace   When using a fireplace, burn only seasoned hardwood if possible. Avoid burning green or wet wood, or softwood such as pine, spruce, and other evergreens; these burn less completely than hardwood, producing more smoke and attendant particles and also (especially in the case of softwoods) producing a tarry, flammable residue called creosote that lines chimneys and poses a fire hazard. Never burn scraps of preservative-treated, painted, or finished wood, or manufactured wood products such as plywood and particleboard; the smoke they produce contains toxic fumes.

To prevent smoke from escaping into a room, build fires well back from the fireplace opening and elevate fuel a few inches above the fireplace floor by placing it on a grate or andirons. Raising the fuel allows air to flow beneath it—enhancing combustion—and also reduces the amount of space above the flames—in effect, bringing the source of the smoke closer to the chimney. Be sure the damper is open fully when you light a fire and while it takes hold; only when the fire is well under way should you close the damper in small increments with a poker to control the blaze and economize on fuel.

Before lighting a fire it may help to ignite a few sheets of rolled newspaper in the fireplace and allow them to burn completely. This warms the air in the fireplace and helps initiate an updraft that will draw smoke up the chimney instead of allowing it to drift into the room. Use a shovel to control large embers of paper; they have a tendency to rise on updrafts and sail dangerously about. Light the fire by igniting kindling near the base. If smoking occurs even when following these procedures, try opening a window near the fireplace (or at least in the same room) a few inches to admit more air. This should strengthen the updraft.

If smoking persists, experiment with lowering the height of the fireplace opening. Use a strip of scrap aluminum siding about fifteen inches deep or a piece of plywood covered with aluminum foil. While a fire is burning, hold the panel against the front of the fireplace above the opening and then gradually lower it until the smoke is contained. Mark the panel to record the point at which this occurred.

Replace the panel with a permanently installed deflector or fireplace hood of the required size. These are available through fireplace and woodstove dealers. Another solution is to install folding glass doors across the opening. When closed, these doors seal the fireplace but allow air for combustion to enter through vents at the bottom. Heat is transmitted into the room through the glass.

A smoking fireplace can also result from outdoor conditions. Downdrafts—cold air spilling down the chimney—may cause smoke to enter a room in puffs. This is often unavoidable during windy weather, and if the problem is only occasional it can be ignored. However, if downdrafts are frequent, the chimney may require modification by raising it or installing a chimney cap (see "Chimneys," below). For advice on installing a new, freestanding fireplace, see "Healthy Heating Systems," beginning on page 102.

Chimneys

Masonry chimneys should be inspected and cleaned annually if they are used regularly by wood-burning fireplaces. If their use is only sporadic—say ten or fifteen fires a year—inspection and cleaning need take place only at intervals of two or three years. A chimney used exclusively by a furnace rarely needs inspection, but such chimneys should be checked promptly if smoke or combustion odor can be smelled indoors, especially in upper rooms.

With chimneys used for wood fires, the problem—as was mentioned earlier (see "Fireplaces")—is creosote. This flammable encrustation that lines chimneys hardens like glass when it cools and rather quickly becomes difficult to remove. For this reason, having chimneys cleaned early in the spring is a better idea than waiting until late summer or fall. Besides being a fire hazard, creosote itself is both a combustion pollutant and a cause of excessive smoke. Thick deposits can even crack masonry. The EPA recommends cleaning chimneys whenever creosote and other deposits are a quarter-inch thick; other safety authorities recommend cleaning when the thickness of deposits is only an eighth of an inch.

To have a chimney inspected and cleaned, hire a professional chimney sweep. The cost should be around $150. While most intrepid homeowners who are unafraid of heights can perform cleaning

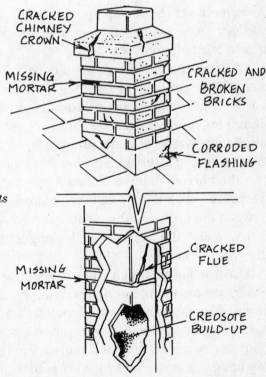

Illustration 27. Chimney faults

tasks—but not inspections—the job requires special tools and can be unimaginably messy if not done right.

Few states license or certify chimney sweeps. The best way to find a good one is to ask for recommendations from friends and perhaps the local fire department. Be sure the sweep you hire can show proof of workmen's compensation insurance. Many professional sweeps belong to the National Chimney Sweep Guild or the Chimney Safety Institute of America, both of which test and certify members. For state-of-the-art inspection, select a sweep offering video inspection service, in which a video camera is lowered into the chimney. Viewing the result on a television monitor, you and an inspector can plainly determine the condition of the interior and tell whether the chimney is safe to use.

Cleaning should involve attaching a powerful vacuum cleaner to the base of the chimney and brushing the flue with long, flexible brushes. Lowering chains and banging them back and forth inside the chimney is an old-fashioned cleaning method that is not recommended because of the damage it can do to flue tiles and mortar. Other methods to pass up are hoisting a small evergreen tree up the chimney from below and periodically building hotter-than-normal fires to burn off creosote.

Chimney damage, too, should be left to professionals in most cases. If the problem is internal—cracked flue tiles or deteriorated mortar joints—the wisest solution is to have the chimney relined. A poured-in-place cementitious liner is best but is also expensive. The installation processes are proprietary. The most common one involves inserting a tubelike rubber form into the chimney and inflating it to the correct diameter; then a special mix of refractory cement and perlite or another insulative material is pumped into the space between the form and the chimney walls. When the mix hardens, the form is deflated and removed, leaving a liner that should endure for fifty years or more.

Clay tile liners, which come in sections that must be mortared together, are also long-lasting and safe, but installing them usually requires disassembling the chimney and rebuilding it—possibly feasible during a chimney restoration but not something to consider if the chimney is otherwise sound except for the liner. Also, installing clay liner sections correctly so that their joints do not soon crack is tricky even for an experienced fireplace mason, and a liner with cracked joints is unsafe.

If you cannot afford a poured-in-place liner, hire an experienced chimney contractor to install a stainless steel liner carefully selected for use with the type of fuel to be burned. Smoke from different

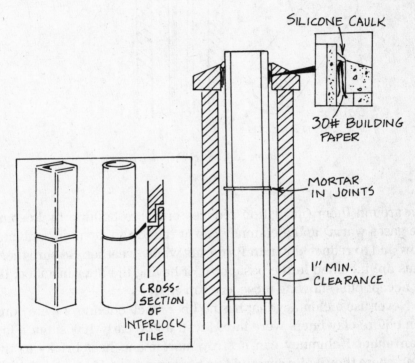

Illustration 28.   Clay chimney tiles

fuels—wood, oil, gas, and coal—contains different compounds, some of which corrode steel alloys more readily than others. Only an experienced and knowledgeable chimney expert should decide the type of liner to use. However, insist that whatever liner is selected meets the requirements of Underwriters Laboratory Standard 1777, indicating that it is adequately designed to withstand the high temperatures (up to 2,100 degrees Fahrenheit) that can occur during chimney fires. (Even so, a steel liner subjected to a chimney fire must be replaced because high heat makes the metal brittle and destroys its corrosion-resisting properties.) Also, steel liners require insula-

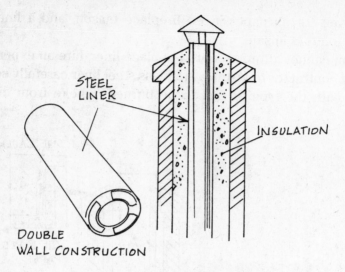

STEEL
LINER

INSULATION

DOUBLE
WALL CONSTRUCTION

Illustration 28A. Steel chimney liner

tion around them (insulation reduces creosote buildup by keeping flue gases warm) and a chimney cap at the top to keep out rain and snow and to reduce downdrafts during windy weather. Be sure these items are included in the basic cost of having the liner installed; do not accept bids showing these as extra charges.

Never use a chimney that has no liner. Such chimneys were common before clay liners were introduced in the early 1900s, but a fire in an unlined chimney can destroy mortar and heat bricks to the point where they ignite surrounding house framing. If you uncover an unused chimney or fireplace during remodeling, have it thoroughly inspected before using it. A walled-up fireplace or chimney may have been intentionally sealed because it smoked or otherwise malfunctioned; early fireplaces often were built by amateurs. Also, never connect differing types of combustion appliances (a wood stove and a gas furnace, for example) to the same flue. Flues connected to differing appliances may be contained in the same *chimney*, but it is dangerous, and a code violation, to combine fuel gases of different appliances in the same flue.

Repairing external chimney damage usually is not as difficult as making interior repairs. Repointing (renewing old mortar) and replacing broken bricks are done as for any brickwork, including fireplace masonry (described earlier—see "Fireplaces"). If you have an

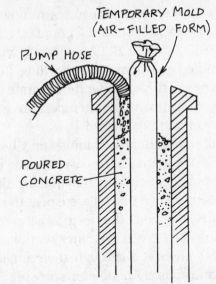

TEMPORARY MOLD
(AIR-FILLED FORM)

PUMP HOSE

Illustration 28B.
Cast-in-place liner

POURED
CONCRETE

old chimney that has been exposed to the elements, it is important to have a mason examine it first to determine whether modern mortar is compatible with the bricks used in its construction. Old mortars (dating from the last century) typically contained only lime and sand and were somewhat flexible; this allowed the softer bricks of that time to shrink and swell due to weather conditions. Modern mortar is made with portland cement instead of lime and is rigid. Using portland cement mortar in normal proportions can cause soft bricks to spall, or flake, which can ruin them. An experienced mason can suggest a mortar formula consisting of a small amount of portland cement supplemented by lime that will work effectively. One such formula is one part portland cement to three parts lime and twelve to twenty parts sand; adding more sand creates a weaker, more flexible mortar.

Another common exterior problem with chimneys is a cracked chimney crown, the concrete cap surrounding the flue opening at the top of the chimney. Very often this cap cracks due to incorrect installation of clay liner sections. Clay expands and contracts naturally with heat and cold; if the seam between the top liner section and the crown is sealed with mortar, such movement will crack either the crown or the liner, and usually sooner rather than later. Repairing a cracked crown sometimes can be accomplished by enlarging the crack with a chisel and filling the gap with mortar or caulking compound. To replace a broken crown requires removing the damaged one, building a wooden form around the chimney top, and pouring a new crown using fresh concrete. Before pouring, wrap a folded strip of thirty-pound building paper around the flue, flush with the intended top of the crown, to permanently separate it from the concrete. Cover the paper strip with high-temperature silicone caulking compound after the concrete has cured.

Water damage indoors around a chimney may be caused by deteriorated roof flashing. When this is the case, coating the flashing with asphalt or plastic roofing cement or having the flashing replaced should cure the problem. When this fails, suspect the chimney itself if it is clay-lined. Corrosive compounds in water condensing from smoke can erode mortar between clay liner sections. When this happens, sections absorb moisture through their unglazed edges and eventually can become saturated. Moisture passing from the soaked liners to adjacent walls and framing then causes staining and paint blistering, resembling damage caused by a leak.

Of course, a chimney also plays a part in how well a fireplace or other combustion appliance behaves. Chimneys draw smoke upward because they contain hot air, which rises naturally, and because outdoor air blowing across the chimney opening creates a partial vacuum that increases the upward movement. For a chimney to capture enough hot air to draw effectively, it must rise at least fifteen feet above the heat source; for outdoor air currents to blow properly across the top, it must extend at least three feet above the highest point where it passes through or lies against the roof and at least two feet above any other potential obstruction within a horizontal radius of ten feet. In addition, the volume of the chimney must bear the correct proportional relationship to the height. Too narrow a chimney (or excessively rough interior surfaces) restricts airflow by increasing friction; too wide a chimney causes the upward velocity of smoke to dissipate and also results in premature cooling that exacerbates the problem and promotes creosote formation.

Little can be done with an ill-designed chimney except to lengthen it at the top by adding courses of additional brick and a section of flue liner. However, do not attempt such renovation without consulting a chimney expert. If downdrafts enter the room during any but the stormiest weather conditions, make sure no foliage overhangs the chimney or interferes with the air flowing around it. Trim back foliage to at least ten feet from the chimney. Also consider installing a metal or masonry chimney cap above the opening to deflect excessive breezes, but only if the problem is severe.

If the smell of stale smoke fills a room hours after a fire has gone out, or if a fire doesn't draw well to begin with, suspect a backdraft. Unlike a downdraft, which results simply from cool air descending naturally into the chimney, a backdraft is the result of an air shortage indoors. This creates a partial vacuum that actually reverses some or all of the flow of air up the chimney.

Backdrafts are more common in well-insulated, weathertight houses than in older, "leakier" homes. When they occur during the use of a fireplace, opening a nearby window a few inches can help—but make sure the window is on the side of the house against which any prevailing wind is blowing. If you open a window on the opposite side, the effect will be to increase the partial vacuum indoors by sucking additional air out; that, in turn, will worsen the backdraft. (For more on backdrafts, see the chapter "Improving Ventilation.")

Woodstoves

Residential woodstoves are among the nation's largest air polluters, and woodsmoke is the leading cause of air pollution in winter in most western cities, according to the EPA. Since 1988, woodstoves sold in the United States have been regulated by performance standards established by the EPA that place strict limits on the amount of pollutants stoves can emit. (Automobiles are the only other consumer product governed by EPA regulations.) But despite the standards, stoves sold in the United States until as late as July 1, 1992, may legally spew pollutants indoors and out at higher than the regulation limit.

Conventional preregulation woodstoves often have only a single combustion chamber (cavity in which wood or other fuel is burned). These stoves are not airtight; their heating efficiency is lower than that of later, airtight models, and they are potentially dangerous sources of combustion products of all kinds. Airtight woodstoves maintain lower temperatures for efficiency and have either a catalytic combustor—a device that removes pollutants from smoke—or else a secondary combustion chamber that collects gases escaping

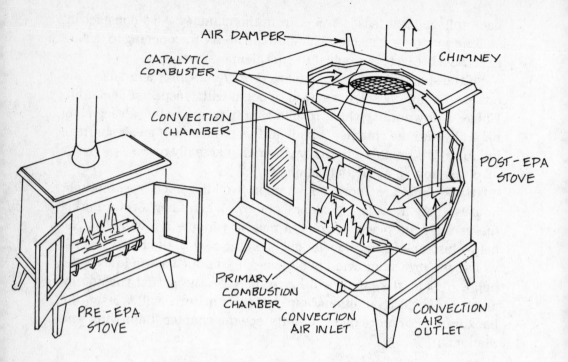

AIR DAMPER

CHIMNEY

CATALYTIC
COMBUSTER

CONVECTION
CHAMBER

POST-EPA
STOVE

PRIMARY
COMBUSTION
CHAMBER

PRE-EPA
STOVE

CONVECTION
AIR INLET

CONVECTION
AIR
OUTLET

Illustration 30. Pre- and post-EPA woodstoves

from the primary chamber and reburns them at a higher temperature. Neither of these techniques completely eliminates contaminants; even the EPA regulations permit pollution levels of 4.1 grams of combustion particles per hour from stoves with catalytic combustors and 7.5 grams per hour from airtight stoves without them. (Pellet stoves, which burn special fuel made of compressed wood and other biomass products, emit combustion particles at rates often less than 1 gram per hour. These are described in more detail in "Healthy Heating Systems," beginning on page 102.

Proper installation is crucial to fire safety when you are dealing with a woodstove (see the chapter "Protecting Your Home from Fire"), but for controlling pollution the crucial element is the stovepipe and its connections. Have your stove and stovepipe installed by reliable professional installers—most states require permits and strict adherence to local building codes. A woodstove should not use an unlined chimney for exhaust; nor should it share a flue with a gas- or oil-fired furnace or other appliance. Unless the

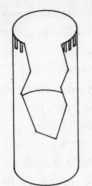

Illustration 31. Stovepipes

SINGLE-WALL   DOUBLE-WALL

chimney is lined and the stove is installed to replace the fireplace, it should be connected to a vertical stovepipe labeled "class A," or "all fuel," or "solid fuel." This is a stainless steel, double-wall stovepipe and is usually required by local fire codes. Single-wall stovepipes rust or corrode easily and permit combustion gases to condense, increasing the formation of creosote.

Sections of single-wall stovepipe can be used to connect the stove to the double-wall stovepipe, but the entire assembly should be as short and as straight as possible to limit creosote and at the same time provide for maximum updraft. Connecting pipes up to about eight feet in length are common; these usually are comprised of a vertical and a gently sloping run joined by an elbow. The length of the sloping run should be less than three-quarters the length of the vertical run and should angle upward toward the vertical double-wall stovepipe at least one-quarter inch per foot along its length. Some installers and owners mistakenly join stovepipe sections upside down; the crimped ends of pipes must point downward, toward the stove, otherwise smoke and creosote can leak from the joints.

The diameter of stovepipes also is important. The correct diameter usually is equal to that of the stove's flue. If the stovepipe is too small for the stove, smoke will leak from the joint and restrict exhaust air flow so that the stove fails to get enough air for good combustion. If the stovepipe is too large, insufficient updraft may result, smoke may not rise quickly, and creosote buildup may be excessive.

Frequently inspect the thimble—the connection between the single-wall stovepipe and the double-wall pipe serving as the chimney. Like all stovepipe joints, this one should be tight and not leak smoke. You should not be able to move the stovepipe assembly more than two inches horizontally.

Clean stovepipes, or have them cleaned by a certified chimney sweep, at least once a year—more often if you use the stove frequently. Monthly cleaning may be necessary in winter, especially if the horizontal pipe section is long, which cools the smoke, or if fuel consists largely of pine or other softwood that burns at a relatively low temperature. Operate a woodstove according to the manufacturer's directions; many of the methods described earlier under "Fireplaces" and "Chimneys" also apply.

Unvented Kerosene Heaters

More than any other combustion appliances, unvented kerosene heaters can be killers. Many community fire and safety codes no longer permit their use. And no wonder: without venting to the outside, all combustion products—including carbon monoxide—produced by these appliances are released indoors, and oxygen in the inside air is consumed. Heater manufacturers recommend opening

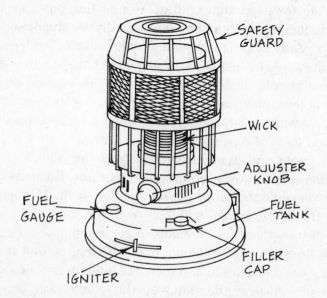

Illustration 32. Portable kerosene heater

doors to any room being heated so that air supplying the heater mixes with air from the rest of the house. However, this usually is inadequate. For sufficient ventilation, fresh outside air should be admitted to the heated room by opening a standard-size window about two inches for every 11,000 Btu of heat for which the heater is rated. A rating label should be attached to the heater.

When properly adjusted, kerosene heaters can burn fuel with almost 100 percent efficiency, far greater than the 60 to 90 percent efficiency typical of oil-burning residential furnaces. However, when they are badly adjusted or using the wrong wick or a wick that is damaged, intolerable amounts of pollutants can quickly be generated.

Fire hazards are another serious concern when using a kerosene heater. For more on this subject, see the chapter "Protecting Your Home from Fire."

Gas Kitchen Ranges

The constant unvented burning of a kitchen range pilot flame can increase levels of combustion products in the air, and igniting burners can raise levels even higher. But while this may be important to owners of newer, tightly sealed houses in which indoor pollutant levels can increase significantly without adequate ventilation, it is worth noting that, so far, evidence that gas ranges contribute substantially to indoor pollution is limited. According to EPA figures, research does show an increase in respiratory ailments among children under two years of age who live in homes with gas ranges compared to children in the same age group who do not. But no supportable evidence exists showing the same effect among adults.

Nevertheless, check the shape and color of burner flames on gas kitchen ranges, or have them checked by a gas company technician, to make certain they are adjusted for maximum combustion efficiency. A proper flame burns steadily and quietly, and displays sharply defined blue cones about half to three-quarters of an inch in height.

Too much air produces a noisy, unsteady flame that may not completely encircle the burner and can also cause hard starting. Besides wasting gas by releasing it unburned into the air, such flames produce increased nitrogen oxides. Too little air results in a weak, sooty flame

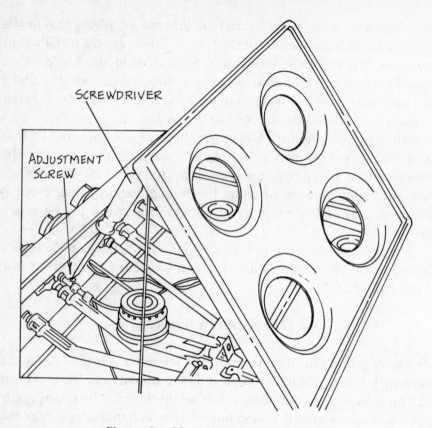

SCREWDRIVER

ADJUSTMENT
SCREW

Illustration 33. Adjusting burner flame

with no blue cones. This kind of maladjustment is more dangerous, as it increases carbon monoxide levels.

Air shutters on tubes leading to the burners regulate the gas/air mixture of the flame. To adjust a burner flame, lift the cooktop (usually it is hinged at the rear) to expose the shutter. Ignite the burner by turning the control knob to High. Loosen the shutter setscrew and open the shutter by turning it until the burner flame becomes noisy and unsteady, indicating excessive air. Then gradually close the shutter to produce a steady flame with sharp blue cones as described above. Retighten the setscrew. Use extreme caution when working around open flames.

If the range is ignited by a pilot flame—a constantly burning stream of gas—check this, too. A pilot flame should flicker softly but produce no smoke or yellow color. To adjust a pilot flame, turn the adjustment screw on the pilot burner or on the gas supply line. Correct adjustment produces a sharp blue cone one-quarter to three-eighths of an inch high.

To adjust oven burners and pilot flames, usually you must remove the door, oven bottom, and baffle beneath the bottom. To adjust an oven burner, first ignite the flame by turning the oven control knob to High. Observe the flames to determine whether they are receiving too much or too little air; then turn off the burner and let it cool for a few

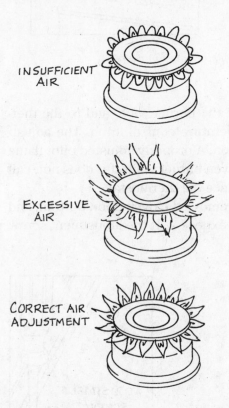

INSUFFICIENT AIR

EXCESSIVE AIR

CORRECT AIR ADJUSTMENT

Illustration 33A.
Correct gas/air adjustment

moments. With the control at Off, find the air shutter at the base of the burner tube; loosen the setscrew securing it, and rotate the shutter to increase or decrease the air opening beneath it. Relight the burner and check the result. Repeat the process until each flame is a steady cone about an inch long with a sharply defined inner cone about half an inch long. Then turn the oven off and reattach the removed parts.

Oven pilot lights are linked to the oven's safety valve, which is

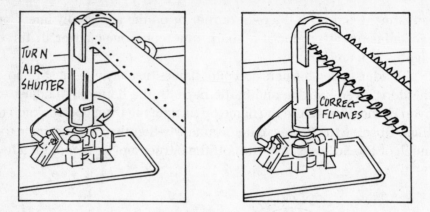

Illustration 33B. Adjusting oven burners

near the air shutter at the base of the burner tube, and to the thermostat, which is behind the temperature control knob. The adjustment screw may be at either location. A properly adjusted pilot flame should increase in size when the oven burner is lit. If it does not, call the gas company to inspect or replace the thermostat.

To adjust an oven pilot flame, remove the oven door, bottom, and baffle as was mentioned earlier to expose it. If the adjustment screw

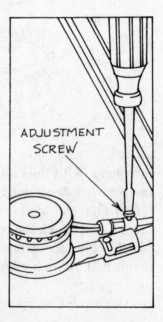

Illustration 34. Adjusting pilot flame

TOO LOW   CORRECT   TOO HIGH

Illustration 34A. Correct pilot flames

is located on the safety valve, turn the screw slowly with a screw-driver to produce a steady blue flame about a quarter of an inch high with a yellow tip. If the adjustment screw is at the thermostat, it will be visible when you pull off the temperature control knob; usually the screw is marked "P" or "Pilot." Turn it as was described above to produce the correct flame. Reinstall the knob.

If range-top or oven burners emit uneven flames or if some ports appear clogged, gently ream the ports with a wire or needle to open them. Use the same technique to open a clogged pilot jet. Range-top burners can be removed easily for washing; removing an oven burner, and servicing electronic ignition and sealed burners are jobs for a professional service technician.

Gas ranges must be vented by a range hood that delivers exhaust gases to the outdoors. Updraft hoods (located above the range) should draw off exhaust at the rate of about 300 cubic feet per minute (cfm) and be located twenty-four to thirty inches above the range surface. Downdraft vents (located on the cooktop) should remove

Illustration 34B.
Adjusting oven pilot light

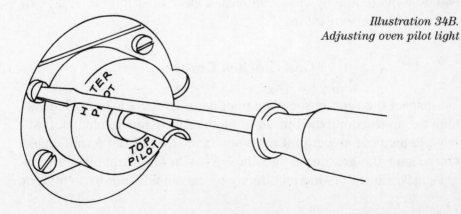

exhaust at about 400 cfm to overcome the natural tendency of gases to rise. To inspect a range hood, make sure that all connections are tight and that the filter is clean. Also inspect the vent opening on the outside of the house to make sure it is not blocked. For advice and instructions on installing a new or replacement range hood, see the chapter "Improving Ventilation."

Gas ranges must be adjusted to accommodate either liquid propane (LP) or natural gas. If you want to change the type of gas your range uses, call the gas company.

Call the gas company also for any repairs involving the main gas supply line, burner control valves, oven thermostat, and oven safety valve. Damage or incorrect repairs to these parts can create hazardous gas leaks. (CAUTION: Never pull any gas-fired appliance away from a wall unless you are certain the gas line is flexible; doing so can damage rigid pipe connections, causing a dangerous gas leak.)

Gas Water Heaters

Inspect gas water heaters as you would gas furnaces or kitchen ranges. Unobstructed venting to the outside is crucial to prevent backdrafts and the buildup of indoor pollutants. Also, the heater's burner flame and pilot (if there is one) must be kept properly adjusted. Water heaters must be located as close as possible to the chimney to reduce the distance exhaust gases must travel. Gas water heaters should not be located in confined spaces like closets, especially those opening into bedrooms or bathrooms, where occupants could be overcome by either unburned gas or carbon monoxide, or be injured by an explosion.

Gas Clothes Dryers

Inspect the vent, burner, and pilot flame (if there is one), and vacuum the area around the burner often to remove inevitable lint. (CAUTION: Remember, never pull a gas appliance away from a wall unless you're sure the gas line is flexible.) As with all gas appliances, exhaust gases must be ducted directly to the outside, not to an attic or

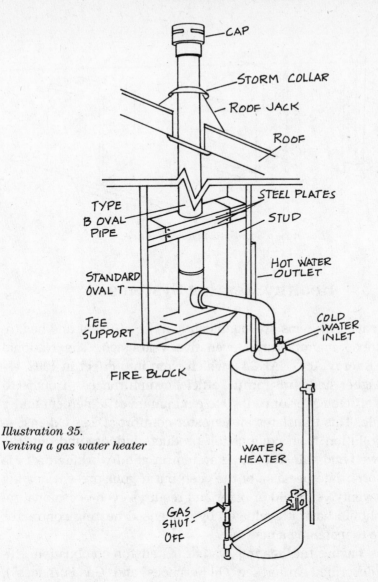

CAP

STORM COLLAR

ROOF JACK

ROOF

STEEL PLATES

TYPE
B OVAL
PIPE

STUD

HOT WATER
OUTLET

STANDARD
OVAL T

TEE
SUPPORT

COLD
WATER
INLET

FIRE BLOCK

WATER
HEATER

Illustration 35.
Venting a gas water heater

GAS
SHUT-
OFF

basement crawl space. Avoid using devices that permit dryer exhaust to be released into the house to add or conserve heat. These allow combustion pollutants to enter indoor air as well, and greatly raise humidity levels. Dryers, of course, remove moisture from clothes by evaporating it with heat, which is then exhausted. Venting any dryer—gas or electric—into a house merely transfers the moisture removed from the clothes into the indoor air.

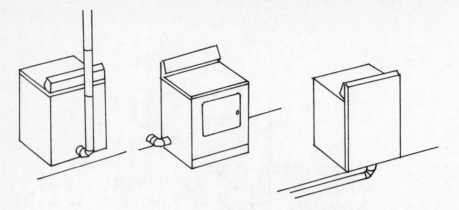

Illustration 36. Venting a clothes dryer

Healthy Heating Systems

New heating systems are significantly more efficient and healthful than those manufactured even five years ago. The National Appliance Energy Act (NAECA), which went into effect in 1992, established higher standards for fuel efficiency in furnaces—from 50 to 60 percent efficiency prior to 1979 to a minimum of 78 percent under the new rule. This translates into greater comfort at less expense—and less pollution from combustion products indoors and out. Replacing a well-maintained heating system in good working order is seldom appropriate because of the cost; but if your present heating system is twenty years old or older and requires even occasional repair, you should have it evaluated by a licensed heating contractor with an eye to installing a new one.

Besides having the heating system tested for combustion efficiency (as described earlier, see "Oil Furnaces" and "Gas Furnaces"), find out whether the furnace is correctly sized for your home in its present condition. Before the growth of energy costs beginning in the 1970s, heating systems typically were deliberately oversized; for builders, this was an easy and practically foolproof way of avoiding the risk of installing a system that was too small, and it allowed for future remodeling that might enlarge the house.

But an oversized heating system cycles on and off more frequently than does a properly sized furnace. This can produce uncomfortable and unhealthful temperature swings, and it also results in less efficient combustion—firing temperatures inside a furnace are lower in the first few minutes of each burn because the combustion chamber and other components have had a chance to become cool.

Take into account the expected life span of your equipment. Boilers, because their parts are subject to relatively low stress, generally are expected to last thirty-five years if they are well maintained. Oil and gas furnaces usually wear out after twenty to twenty-five years, and heat pumps typically last only ten to fifteen years. Warning signs that a heating system is about to expire are increased fuel consumption, burner problems (these often cost several hundred dollars to repair; investing the money in a new system may make more sense), excessive sludge in boiler water, and cracks in a furnace's heat exchanger.

With all of this information in mind, consider the payback period of installing a new system—the length of time required for the system's parts and installation costs to pay for themselves with fuel savings and reduced maintenance charges. A payback period of five to seven years is considered reasonable, provided you plan to live in the house that long or can recoup the remaining cost when you sell. Upgrading to a new heating system can increase a home's value if the present heating system is twenty-five years old or older; replacing a newer system usually does not add value. Payback is a moot issue, however, if health is at stake.

What kind of new system should you consider? New furnace technology has developed along three lines. Most important from the standpoint of emissions are condensing furnaces and sealed-combustion furnaces.

Condensing furnaces contain two heat exchangers instead of one. While the primary exchanger functions to warm air or water for heating, the second processes combustion gases leaving the furnace in the form of exhaust. Heat from these, too, which ordinarily would escape up the chimney, is extracted and used for the same purpose as that flowing through the primary exchanger. The result is that the temperature of combustion gases in the chimney is so low that any water vapor they contain condenses out and is drained away by a pipe.

Cooler combustion gases permit condensing furnaces to exhaust directly through a nearby wall, with no chimney required. In fact, the exhaust flue itself usually can be of plastic pipe. This can be very con-

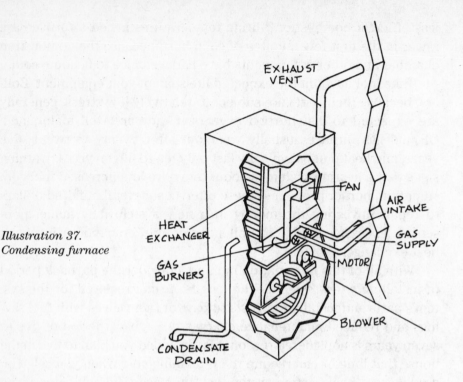

Illustration 37.
Condensing furnace

venient in remodeling and new construction. However, when replacing an old furnace with a condensing model, venting the new furnace into the chimney used by the old one may require relining the existing chimney to withstand increased moisture and acids (see "Chimneys," above).

Sealed-combustion furnaces are a boon to owners of tightly built houses, where backdrafting can be a problem. Intake air to a sealed-combustion furnace is supplied directly by a pipe that connects the unit to the outdoors; therefore the furnace uses no indoor air for combustion and so cannot contribute to creating backdrafts. Further, exhaust air is forced from the furnace to the outdoors by means of a powerful fan.

Coupled with a condensing furnace, sealed-combustion technology assures the safest possible home heating apart from nonfuel systems (for example, electric and solar). But existing conventional furnaces can be retrofitted with sealed-combustion features to create a system that is nearly as safe. Retrofit components consist of intake and vent ducts plus either a power burner (for oil-fired furnaces) that creates sufficient pressure to force combustion gases through the vent duct to the outside, or a suction-inducing fan that instead draws exhaust through the vent.

The third important feature of newer furnaces is a variable-speed blower. While not as important in reducing combustion gases as the other two technologies, this device adjusts airflow through ducts to match the heating needs of the house. Naturally, this increases heating efficiency, but it also affects indoor humidity levels for the better and helps maintain constant temperatures. Both are important in preventing colds and other respiratory ailments that can be brought on by harsh or uneven indoor heating.

Installing a New Fireplace

Modern, prefabricated metal fireplaces are safer, operate more efficiently, and are far less expensive than their hand-built masonry forebears. They have virtually replaced traditional fireplaces in new construction and remodeling. Besides low cost, their chief advantage is versatility; they can be installed practically anywhere in a house.

Most prefabricated fireplaces have double-wall construction— typically, an inner wall of refractory material resembling firebrick and an outer wall of heavy-gauge steel. Between the walls is an insulating airspace; this allows such fireplaces to be installed as close as five eighths of an inch to flammable materials such as walls and house framing. For this reason, such fireplaces are referred to as "zero-clearance" models. Generally, a multiwalled steel chimney having similar insulating qualities also is installed.

Gas-burning zero-clearance fireplaces are the cleanest, albeit the least "realistic" in appearance, compared to the wood-burning kind. However, newer models use improved burner technology to produce a yellow flame instead of the blue flame characteristic of gas combustion; and even the ceramic "logs" in some designs now look genuine.

A major development is the direct-vent gas fireplace that does not require a chimney. A special double-wall flue is used instead; the flue exhausts combustion gases through a small pipe, or inner flue, at the same time that outdoor air enters to feed the flames through the larger flue surrounding it. The incoming air both cools the exiting combustion gases and protects surrounding walls from heat that

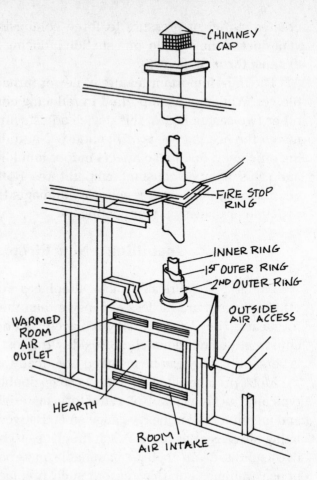

Illustration 38.
Zero-clearance fireplace

does escape. Indoor air is not used for combustion; therefore, there is no risk of backdrafts unless the fireplace doors (which are clear glass) are open.

Wood-burning zero-clearance fireplaces borrow much from contemporary woodstove technology, often including high standards of emissions control. While fireplace emissions are not regulated by the EPA or other authorities, many of these new fireplaces meet or exceed current EPA woodstove standards (for more on these standards, see "Woodstoves," earlier in this chapter).

The latest wood-burning fireplace technology involves positioning a baffle, or a grid of perforated tubes, near the top of the firebox. This creates a secondary combustion chamber rather like the ones

106

on many airtight woodstoves. Smoke from burning fuel rises from the fire toward the front of the fireplace (which, like the gas model described earlier, is sealed by glass doors) and then upward, where it circulates around the baffle and enters the secondary chamber.

Air in the baffle, heated by the fire, issues under pressure from the perforations, igniting the circulating smoke. This reduces their combustion products further and increases the amount of heat the fireplace delivers.

Other wood-burning zero-clearance fireplaces feature catalytic converters, which cause products in combustion gases to burn at lower temperatures, thereby ensuring that they are consumed before rising up the chimney.

Expect to pay anywhere from $500 to $1,500 for a zero-clearance gas or wood-burning fireplace, and extra for chimney sections. While they are not too difficult to install for an amateur with carpentry skills, following the manufacturer's instructions carefully is a must. In addition, installations must adhere to local building code regulations and usually require a permit and an inspection.

Pellet Stoves: A New Woodstove Technology

Woodstoves have undergone a revolution in combustion cleanliness and efficiency since federal emissions regulations for them were enacted in 1988 (for more on these, see "Woodstoves," earlier in this chapter). The newest permutation is the pellet stove, named for the fuel it consumes—pressed nuggets made of refined sawdust, wood shavings, nutshells, and similar cellulose materials. Unlike firewood, pellet fuel contains virtually no water; therefore it burns hotter and produces fewer combustion products (little ash and no creosote, for example). Whereas the strictest current EPA regulations limit conventional woodstoves to particle-emission rates of 4.1 grams per hour, pellet stoves often emit fewer than 1 gram per hour.

Pellet stoves incorporate highly technical, electronically controlled combustion methods but provide reliable, efficient heat in return. Pellets are poured into a hopper, where an auger or another mechanism feeds them at an adjustable rate into a ceramic-lined combustion chamber. Lighting the stove is done by hand (with a match) or by means of an automatic igniter. An electric blower then

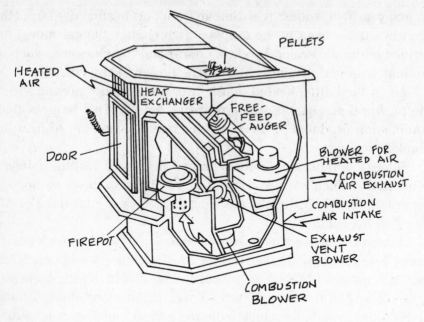

PELLETS

HEATED AIR

HEAT EXCHANGER

FREE-FEED AUGER

DOOR

BLOWER FOR HEATED AIR

COMBUSTION AIR EXHAUST

COMBUSTION AIR INTAKE

FIREPOT

EXHAUST VENT BLOWER

COMBUSTION BLOWER

Illustration 39. Pellet stove

draws air from outdoors into the chamber, enhancing the flame. Another blower exhausts combustion gases to the chimney, which is usually smaller than the flue needed for a conventional woodstove.

The rate at which pellets are fed into the combustion chamber regulates the size of the fire. If the blower is turned off, the fire will go out. During operation, indoor air is drawn into part of the stove and across a heat exchanger. The exchanger transfers the heat inside the combustion chamber to the air passing over it, just as with a central-heating furnace and forced-air distribution system.

Unlike conventional woodstoves, pellet stoves heat by convection, not radiation; therefore, their surfaces rarely get too hot to touch—good news especially for households with children. Typically, pellet stoves require three inches of clearance at the sides and one inch at the rear; they must rest on a noncombustible floor surface at least three-eighths of an inch thick.

Drawbacks to pellet stoves are price (most stoves cost between $1,500 and $2,500), the availability of pellets (they are harder to obtain than firewood in some areas), and the need for frequent servicing due to their complexity. Pellet stoves require electricity to operate;

something to consider if you live where power outages are routine. Also, few pellet stoves display as inviting a flame as do more conventional woodstoves. If you plan to enjoy watching a fire inside the stove, select a model especially designed for viewing. These have a large window and often ceramic "logs" that help simulate the appearance of a woodstove or fireplace fire.

Biological Contaminants

What Are They?

ALMOST EVERYONE HAS SNEEZED OR EXPERIENCED NASAL and respiratory discomfort in the presence of household dust. The chief ingredients of such dust are organic materials, microscopic organisms, and their by-products. Air quality investigators refer to all of them as biological contaminants. Some people are severely affected by these contaminants and develop asthma and other illnesses, sometimes with serious consequences.

A breakdown of household dust typically reveals skin flakes and hair (from humans and pets), molds, insect parts and droppings, and

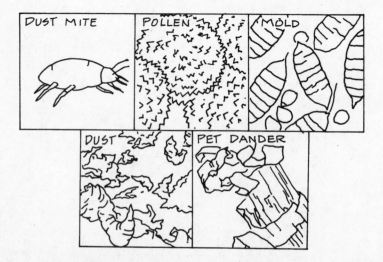

Illustration 40. Biological contaminants

particles shed by fabrics, upholstery, carpet, and other materials used in household products. All of these particles can harbor viruses, bacteria, protozoa, and other microorganisms. Among the most irritating substances, according to allergy experts, are molds (including fungi) and the feces of the nearly ubiquitous dust mite, a microscopic creature related to spiders. Both thrive in conditions of poor air circulation and high humidity.

How Bad Is the Problem?

Some biological contaminants associated with indoor air are either toxic or else create toxins and can cause infectious disease. The *Legionella pneumophila* bacterium, for example, causes Legionnaire's disease and its lesser relative, Pontiac fever, plus an estimated 13 percent of all pneumonia seen in the United States. While it is found outdoors to some degree, *Legionella* spawns readily in residential forced-air heating systems, humidifiers, water systems, vaporizers, hot tubs, and as a result of flooding in basements and other household areas.

Most indoor biological contaminants produce allergic reactions. These range from allergic rhinitis (hay fever), which carries the risk of secondary sinus infection, to so-called humidifier fever, hypersensitivity pneumonitis, and bronchopulmonary aspergillosis. Allergic reactions occur in people whose immune systems erroneously interpret an otherwise harmless agent—for example, animal hair—as dangerous to the body. The immune system responds by producing antibodies that attack the allergen; during the process a chemical called histamine is released into the tissues of the skin, eyes, nose, throat, and lungs. Histamine produces the itching, welts, watering, and congestion that are the symptoms of most allergies. According to figures published by the EPA, 15 percent of the population in the United States suffers from hay fever symptoms brought on by allergic reactions; at least 3 to 5 percent suffer asthma instead, and that percentage appears to be increasing.

Humidifier fever and hypersensitivity pneumonitis produce different symptoms. These maladies are characterized by flulike episodes of chills, fever, muscle aches, and fatigue. Hypersensitivity pneumonitis also produces a dry cough, shortness of breath, and chest tightness. Over time, scarring of the lungs can occur, leading to pulmonary failure and death in severe cases. The disease can result from repeated individual exposures to high levels of allergy-producing agents, biological or otherwise, or from continuous exposure to low levels. In the latter case, the symptoms described may not appear; nevertheless, scarring of lung tissue may go on unnoticed until it is too late for it to be reversed.

Allergic bronchopulmonary aspergillosis is a pneumonia caused by an allergic reaction to a common fungus, *Aspergillus fumigatus*, whose minute spores are airborne and so small they pass easily through the upper respiratory tract (where larger allergens, such as most pollens, are trapped) and penetrate deeply into the lungs.

Certain individuals—perhaps 20 percent of the population, according to EPA figures—may be genetically predisposed to allergies. In these people, prolonged exposure to potential biological allergens can create sensitivity to them where none existed before.

Testing

Two conditions foster most biological growth: nutrients, and constant moisture with poor air circulation. Dusty areas of all kinds contain abundant nutrients on which biological pollutants can thrive and quickly multiply; appliances that add moisture to the air, combined with inadequate ventilation to remove that moisture, complete the equation.

Many symptoms occupants of a home may suffer due to biological contamination are similar to those caused by other indoor pollutants—namely, combustion products and volatile organic compounds (VOCs, which are fully described in the next chapter). But biological contaminants are the primary factor to consider if you can answer yes to any of the following questions about your home:

- Do areas of the house feel humid? Do bathrooms, kitchens, and laundry rooms remain moist for long periods after use?
- Does moisture form on windows or other surfaces?
- Are areas of the house typically very hot or very cold?
- Has the house suffered water damage within the past twelve months?

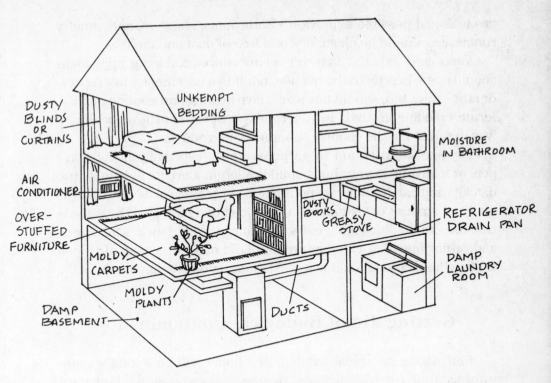

DUSTY
BLINDS
OR
CURTAINS

UNKEMPT
BEDDING

MOISTURE
IN BATHROOM

AIR
CONDITIONER

OVER-
STUFFED
FURNITURE

DUSTY
BOOKS
GREASY
STOVE

REFRIGERATOR
DRAIN PAN

MOLDY
CARPETS

DAMP
LAUNDRY
ROOM

MOLDY
PLANTS

DAMP
BASEMENT

DUCTS

Illustration 41. Contaminant locations

- Is the basement or crawl space damp?
- Are areas of mold or mildew (a black, powdery residue) visible?
- Does any part of the house, including the basement, have a musty odor?
- Are air conditioners, humidifiers, and dehumidifiers cleaned regularly?
- Do houseplants appear moldy?
- Have you seen cockroaches, rodents, or other pests indoors?
- Do you keep pets?

To discover biological contaminants and conditions that promote their growth, tour the house, including the basement and attic, to check for the above conditions. Be sure to check for moisture beneath refrigerators and for greasy dust and other debris beneath and behind ranges and kitchen cabinets and on range hoods. (CAUTION: Do not pull any gas appliance away from the wall unless you are certain it is connected with a flexible gas line. Doing so could cause a

113

gas leak and possible explosion.) Bathrooms, showers, and laundry rooms also should be clean, dry, and free of dust and mold.

Vents and exhaust fans for all moisture-producing appliances should open directly to the outside, not into a basement, crawl space, or attic. They, too, should be clean. Have heating and cooling ducts—fertile breeding grounds for microbes—inspected professionally, as was described in the chapter "Combustion Products."

Pay special attention to carpeted areas, especially if you have pets or if anyone in your household is allergic. Carpets hide moisture damage and can harbor vast populations of biological agents that virtually no amount of cleaning can permanently eradicate. The same is true for stuffed sofas and chairs, bedding, drapes, fabric wallpapers, and furnishings that can capture and hold dust—for example, open shelves, book and knickknack collections, and venetian blinds.

Getting Rid of Biological Contaminants

Eliminating microbial habitats in a home will go a long way toward getting rid of microbes themselves and related biological agents. Frequent housecleaning combined with removing problem items is often effective; remodeling certain areas and installing equipment to control humidity, improve ventilation, and in some cases filter indoor air also may be desirable. (For a complete discussion of ventilation, see the chapter "Improving Ventilation.")

While cleaning the entire house and keeping it clean is a desirable goal, practically speaking it is seldom achievable. Instead, prioritize large tasks and schedule them for special treatment, and focus regular cleaning efforts on areas where occupants spend the most time. Gradually expand your activities to include less-frequented areas of the house.

Cleaning bedrooms is extremely important; most people spend a third of their time there, and young children may spend even more, especially if they play in their rooms. High levels of dead skin flakes in mattresses and bedding encourage vast populations of dust mites (dry skin cells, in fact, are dust mites' chief source of nutrients). Vacuuming, including mattresses, removes surface dust and the fecal pellets produced by mites—but not the mites themselves. To get rid of as many mites as possible—they can never be totally eradi-

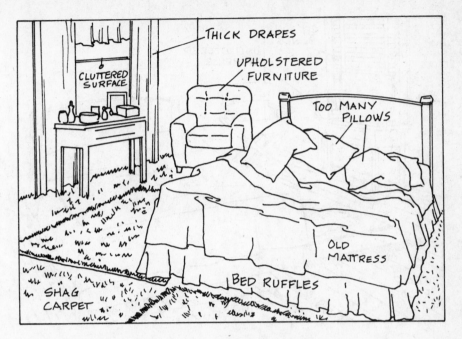

Illustration 42. Dust and clutter are a haven for allergens

cated—lower indoor humidity levels (the subject of indoor humidity is covered farther on in this chapter) and wash bedding in hot water—over 135 degrees Fahrenheit—at least weekly. Encasing mattresses, box springs, and pillows in nonallergenic plastic covers with zippers also is recommended. Each week, wipe the covers with a damp cloth.

Bedroom decor consisting of plentiful fabrics—canopies, dust ruffles, drapes, cushions, and carpeting—harbors both dust and mites. Getting rid of these items, as well as dispensing with feather pillows and down quilts, reduces the potential habitats for microbes and the supply of raw materials for dust—that is, fabric sheddings.

Do away with large collections of books in bedrooms (and elsewhere, if household occupants are allergic). Books generate large amounts of dust and are a breeding ground for molds and fungi, among other microorganisms. Replace ordinary stuffed toys with washable, nonallergenic ones, and launder them in hot water every two or three weeks (weekly, if used often). Naturally, do not allow pets to spend a lot of time in the bedrooms.

Bathrooms and kitchens require regular attention also, primarily

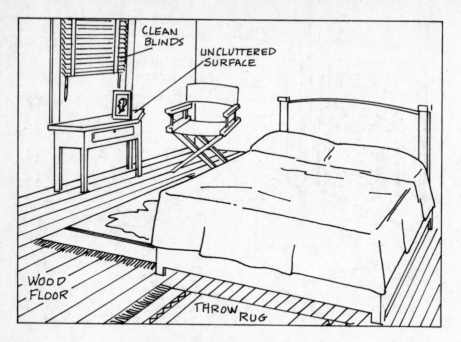

Illustration 42A. Allergy-free bedroom

to keep down mold and bacterial growth. Adequate ventilation usually is one key (see the chapter "Improving Ventilation"); another is washing surfaces to keep them clean.

Wipe bathroom fixtures and floors weekly with bleach or disinfectant. Keep drains free of hair and soapy sludge. Scrub grout between tiles with bleach or a grout-cleaning solution made for the purpose. Replace discolored or deteriorated caulking compound around the bathtub and elsewhere.

Missing caulking or grout allows moisture to accumulate in walls. If you detect moisture damage, which is often signaled by loose ceramic tiles, remodeling is required—both to rid the area of biological contaminants and to halt decay of structural elements. Check for water damage also around toilets and beneath basins. A loose toilet can be a sign that the wax ring sealing the fixture to the waste pipe underneath is damaged, allowing water from flushing to leak into the floor. This problem is fairly common, but if it is not repaired (by having a plumber remove the toilet and install a new ring) serious floor damage can result.

Also, wallpaper in bathrooms showing signs of mold should be replaced with washable, mildew-proof paint. Bathroom floors should never be carpeted; replace carpeting with a hard-surface flooring such as tile or vinyl that can be kept clean and dry. Cover such floors with foam-backed, washable throw rugs for comfort and to avoid slipping.

Follow the same cleaning procedures in kitchens as in bathrooms, but be careful about using disinfectants on surfaces where food is prepared; be sure to rinse residue thoroughly from such surfaces by wiping with a clean, damp cloth—not the cloth used for applying the disinfectant.

Keep food in closed containers and take out garbage as soon as it accumulates. Clean the refrigerator often, especially the drip tray underneath. Stagnant water in the tray is a breeding ground for toxic bacteria as well as for mold, and has been found to be responsible for allergic and disease symptoms in numerous cases investigated by EPA officials and others.

Other regular housecleaning should consist of general dusting and vacuuming. Because these activities temporarily raise levels of airborne contaminants, it is a good idea to wear an inexpensive dust mask when performing them if you are allergic, and to have other allergic household members leave the house. (Incidentally, there is negative value in cleaning a house more than once a week. Doing so only increases airborne allergen levels and does little that weekly cleaning cannot achieve.) As a remodeling project or with new construction, consider installing a central vacuum system. Unlike a portable vacuum cleaner, which can spew out the exhaust small dust particles that are not trapped by the filter bag, a central vacuum draws all dust through a system of pipes installed in the walls to a container located outside the house, usually in the garage or basement.

Keep knickknacks, books, electronic gear, and other dust catchers and dust producers in closed cabinets. Dust surfaces with a damp cloth, not a feather duster or similar tool. Vacuum carpeting, drapes, and upholstered furniture as was described for bedrooms (the same caveats against dust-harboring fabrics and other materials apply), and be sure to regularly wipe roller shades and venetian blinds. (Vertical blinds are easier than horizontal blinds to clean, and they accumulate less dust.)

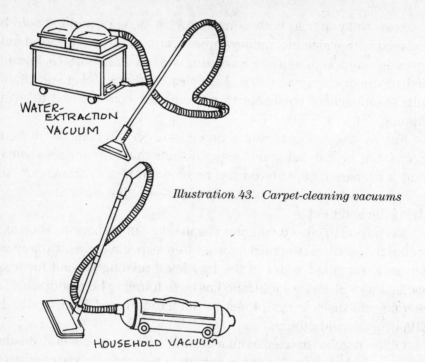

WATER-
EXTRACTION
VACUUM

Illustration 43. Carpet-cleaning vacuums

HOUSEHOLD VACUUM

Cleaning products made with tannic acid for carpeting and upholstered furniture denature (neutralize) allergy-producing compounds in dust mite feces; however, they do not kill the mites. If you choose to use these products, apply them every two or three months in humid areas and every four to six months if you live in a drier region.

Damp-mop wood and other hard-surfaced floors weekly. For regular carpet vacuuming, use a powered cleaning attachment. Equip the vacuum cleaner with microfiltration dust bags that trap small particles. The most effective are HEPA (high-efficiency particulate-arresting) bags. These can trap even pollen, molds, animal dander, and dust mites. Some filter bags also contain disinfectant to retard bacteria growth in the contents.

For periodic deep cleaning of carpets, buy or rent a water-extraction cleaner. Resembling a vacuum cleaner, this appliance injects a shampoo solution under moderate pressure into carpeting and then immediately vacuums it up, together with loosened dirt. The advantage of a water-extraction cleaner is that it uses a minimum

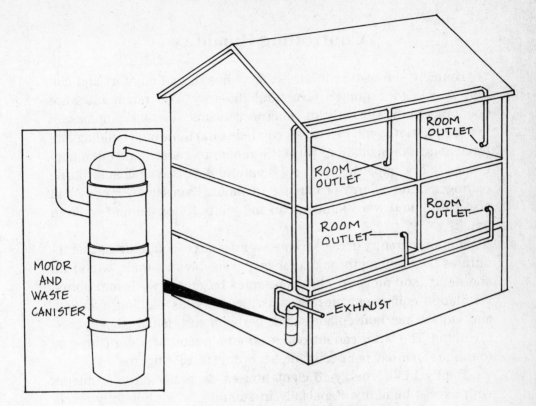

Illustration 43A. Central vacuum cleaning system

amount of water and avoids soaking the carpet and any padding underneath. Because drying often takes place in an hour or so, there is little opportunity for mold and other biological contaminants to grow.

Water-extraction cleaning reduces dust mite populations, but decimating them requires true steam cleaning at sustained temperatures exceeding 130 degrees Fahrenheit. Hire professional carpet cleaners to perform this task; some use a system called "dry steam" that blasts carpeting with moist, heated air at temperatures of over 200 degrees without raising indoor humidity levels. This method can also be used for cleaning drapes and upholstered furniture.

Avoid traditional rug shampoos unless carpets can be dried outdoors. Even then, follow the manufacturer's instructions carefully, as too-rapid drying can result in soap residue that produces an irritating dust.

Controlling Humidity

Humidity—moisture in air—affects health and comfort and can play a role in a home's structural integrity. Too much moisture encourages the growth of microorganisms, creates unpleasant living conditions, and can cause condensation that ruins building materials such as insulation; too little moisture causes mucous membranes to dry, impairing the body's natural defenses against airborne pathogens and allergy-producing irritants. Excessive dryness can also shrink and warp lumber, causing joints to loosen and walls to crack.

Indoor humidity derives from several sources: outdoor air that infiltrates the building through cracks, permeable materials, windows, and doors; soil moisture that penetrates basement walls and floors; ventilation equipment; normal activities such as bathing, cooking, and laundering; houseplants; and, perhaps surprisingly, occupants' breathing. The latter can introduce about a pound of water (three or four pints) per day per occupant, according to EPA figures.

Tightly built, energy-efficient houses generally have problems with excess humidity. Especially in summer, when humidity levels are naturally high, such houses often contain too few escape routes for indoor air. By contrast, older, "leakier" houses tend to have problems with dryness, notably during the winter, when cold, dry outdoor air infiltrates the building and is heated, making it even drier.

Ideal average indoor humidity levels are between 40 and 50 percent in summer and between 30 and 40 percent in winter. These ranges are sufficient to control mite and mold growth, afford reasonable comfort, and eliminate potentially damaging condensation. Notice that these are average levels. Humidity can vary from room to room, and even within a room (because warm air rises, areas near ceilings are often more humid than areas nearer the floor). Keeping overall humidity levels within the ranges specified will reduce the severity of extremes.

You can monitor humidity levels reasonably well with an instrument called a hygrometer. This device contains a spring that expands and contracts as humidity changes, causing a needle to move across a calibrated dial. Many hygrometers resemble wall clocks and are de-

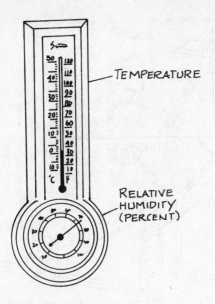

TEMPERATURE

RELATIVE
HUMIDITY
(PERCENT)

Illustration 44. On-the-wall hygrometer

signed for display as well as service. Electronic models with digital readouts are also available. Both kinds can be obtained at department stores, marine supply stores, and some home centers.

To combat seasonal humidity problems, any of several mundane methods may be effective, alone or in combination. Adequate ventilation is always important, whether conditions are too moist or too dry (for a full discussion of this subject, see the chapter "Improving Ventilation"). Simple methods for reducing excessive humidity include running an air conditioner in summer (the process of conditioning air dehumidifies it); drying laundry outdoors or in a clothes dryer vented to the outside; keeping lids on cooking pots when cooking; eliminating large or numerous houseplants; keeping showers short; eliminating basement leaks and dampness (moistureproofing basements is described farther on in this chapter); adding new, tighter-sealing, energy-efficient windows; and weatherstripping outdoor seams to prevent infiltration.

To combat occasional or seasonal dryness, leave exhaust fans off when showering or bathing, simmer a kettle or a pot of water on the stove (add spices or herbs to create a pleasant aroma, but be sure no one in the house is allergic), and keep windows closed and well sealed with weatherstripping.

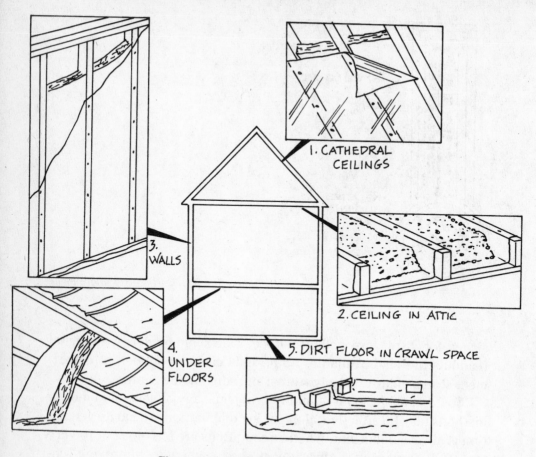

Illustration 45. Vapor barrier locations

For chronic or severe moisture problems, consider installing vapor barriers if they are not already in place. A vapor barrier is any material or coating that reduces the passage of moisture by diffusion through building materials. Thick (four- to six-mil) polyethylene, specially formulated paint, and aluminum foil are commonly used. Nearly all newer houses, including most of those built in the last thirty years, contain vapor barriers. In fact, most building codes require them. Barriers usually are in the form of polyethylene sheeting behind walls, above uninsulated floors and sometimes in top-floor and cathedral ceilings; however, the Kraft-paper and foil facings on fiberglass and other insulation batts also qualify. There are no hard and fast rules about when a vapor barrier is needed in a ceiling. There are too many variables and only limited research has been gathered to date. Most attic moisture problems are due to inadequate ventilation and/or excessive household humidity. However, if you are considering an attic or ceiling vapor barrier, consult a local builder or inspector for advice. If you choose to supplement attic insulation by adding batts on top of existing material, slash the batts' facings with a utility knife before installing them, so that any moisture can escape.

Much of a vapor barrier's effectiveness depends on its ability to bar air, which can contain moisture. In fact, while a vapor barrier that is, say, 75 percent resistant to moisture offers 75 percent effectiveness, unless the material is also virtually 100 percent airtight it will be practically worthless. A hole an inch in diameter (or accumulated small openings totaling the same area) will allow a hundred times as much moisture to pass—contained in air—as a "hole," or nonbarrier surface, of equal size will allow to pass by diffusion.

Homes in northern climates require stronger vapor protection than those in warmer areas because the greater the difference in temperature (and therefore humidity) between indoor and outdoor air, the greater the pressure of the warmer, more humid air to flow toward its cooler, drier counterpart. Where conditions are more nearly equal indoors and out, less pressure occurs. In very warm and humid areas—southern Florida and the Gulf Coast, for example—having no vapor barrier is sometimes better than having one, as the barrier can cause increased pressure from outside that makes it more difficult to maintain drier conditions inside. Where vapor barriers are used in these areas, they generally are installed on the outside of a house to protect the structure from decay.

Ratings, called perms, indicate the ability of moisture to diffuse through, or permeate, a material. The lower the perm rating, the more resistant is the material and therefore the better it functions as a vapor barrier. However, as the chart below shows, nearly all building materials resist moisture to some degree, and in some cases the effect of combining them can be to trap moisture inside walls by allowing it to pass more freely in one direction than the other. To prevent this, building codes generally recommend keeping the normally cooler side of a wall five times more permeable than the warmer side.

MOISTURE RESISTANCE OF COMMON BUILDING MATERIALS

Material	Perm Rating (approx.)
3/4-in. plaster on wood lath	15.0
1/2-in. gypsum wallboard	37.5
1/2-in. plywood	0.4
1-in. foam insulation	1.2
Aluminum foil	0.0
4-mil polyethylene	0.8
6-mil polyethylene	0.6
15-lb. asphalt felt (building paper)	5.6
Kraft-paper insulation facing	0.3
1 coat high-quality acrylic latex primer	8.6
1 coat high-quality acrylic latex semigloss paint	6.6
1 coat latex vapor-retarder paint	0.6
2 coats oil-base enamel on smooth plaster	1.0
1-in. poured concrete	3.2
glass	0.0

As was mentioned above, sealing out air can have more of a moisture-barring effect than installing vapor barrier material itself. In a way this is fortunate, because adding a vapor barrier except during new construction or full-scale remodeling is usually difficult, if not impractical, and sealing out air in existing construction is much easier.

However, closing air gaps must be done thoroughly or the effort will have no effect. Basically, the job consists of caulking or weather-stripping every exterior gap and seam, and sealing openings in inside walls, floors, and ceilings with caulking compound or polyurethane foam—or with weathertight flanges made for the purpose for such items as electrical outlets and recessed light fixtures.

Pay special attention to filling any gaps behind molding. Often a considerable space, which is hidden by baseboard molding, is present between the floor surface and the lower edges of wallboard. Sealing the molding's seams with caulking compound will remedy the situation; if the gaps between the molding and other surfaces are wider than an eighth of an inch, insert foam backing strips, or rods, first and then cover these with caulking. Backer rods are available at

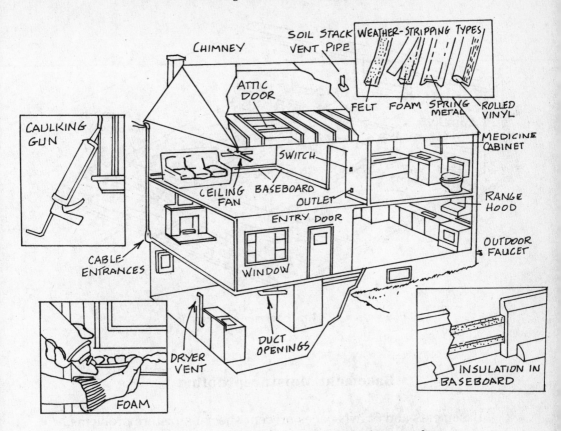

Illustration 46. Caulking and weather stripping to seal gaps

most large home centers and at building supply stores. Another solution is to remove the baseboard, attach adhesive-backed foam weatherstripping to the wall at floor level and above the gap, and then reattach the baseboard so it compresses the foam.

Molding around windows often covers gaps that have been plugged by stuffing them with fiberglass insulation, which does not stop air. Also, when it is compressed, fiberglass no longer insulates. The best solution is to remove the molding and replace the stuffing with polyurethane foam spray insulation, which performs both functions. Buy nonexpanding foam to avoid a mess and the possibility of forcing parts of window frames out of position. Cover the filled seams with weatherstripping tape or adhesive-backed aluminum foil tape for additional draft protection. Reattach the molding.

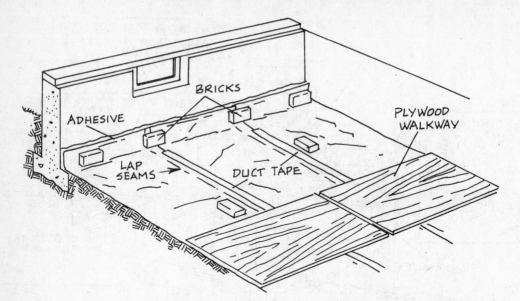

Illustration 47. Vapor barrier over dirt floor

Basement Moistureproofing

Basements and crawl spaces present special moisture problems. It is essential that vapor barriers cover any areas of dirt flooring. The usual method of installing them is to level the floor with a rake, remove sharp stones and debris, and then cover the floor with six-mil plastic sheeting, overlapping the edges by a foot where sheets join. Tape the seams along their entire length with duct tape to seal them. Fold the plastic upward where it meets the walls, and place bricks around the perimeter as anchors. For extra protection, seal the plastic to the walls with construction adhesive. If the floor will be used, install a second layer of plastic or place plywood over heavily trafficked areas. With new construction, consider spreading gravel over the plastic instead. If basement walls are dry-laid stone (this type of construction uses no mortar), you may have to live with the problem unless you can effectively cover the basement ceiling with plastic sheeting to seal the entire basement from the upstairs.

Serious cracks and major leaks in walls and concrete floors should be repaired by a professional basement waterproofer or concrete repair specialist after evaluation by an independent building inspector; all are listed in the telephone book. The material of choice for large cracks is epoxy; however, despite epoxy's abilities, successful permanent repairs depend on locating and remedying the cause of the damage.

Small, isolated cracks and holes, and walls that are only occasionally damp, can often be remedied by amateurs using patching

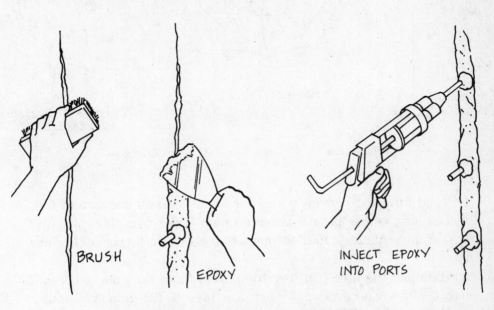

BRUSH

EPOXY

INJECT EPOXY
INTO PORTS

Illustration 48. Filling a basement crack with epoxy

products and waterproofing compounds available at home centers. To repair cracks and small holes, use hydraulic cement. Begin by scouring all loose material from the damaged area with a wire brush. Also, if practical, undercut the sides of the area by chiseling so that the crack or hole is wider toward the inside of the masonry than at the surface. Use a hammer and cold chisel, and wear protective goggles.

Next, dampen the area thoroughly with a sponge (you don't need to do this if water is flowing through the opening), then mix the cement powder with cold water to the consistency of modeling clay, following the instructions on the package.

Fill cracks by applying the cement with a mason's trowel or a putty knife. To fill a hole, put on rubber gloves (cement is mildly caus-

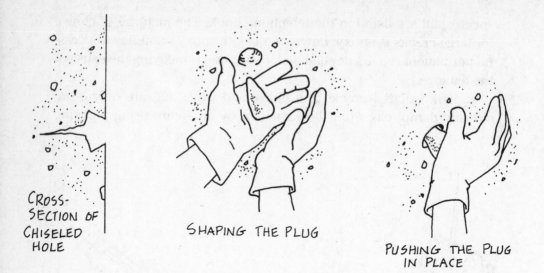

CROSS-
SECTION OF
CHISELED
HOLE

SHAPING THE PLUG

PUSHING THE PLUG
IN PLACE

Illustration 49. Filling a hole with hyraulic cement

tic) and form the cement with your hands into a cone two to four inches long whose base is about an inch wider than the hole. Wait two or three minutes, until the cement feels warm—a sign of hardening.

Dampen the hole if it has dried, then force the cone point-first into the hole and press against it for three to five minutes, until it hardens further. Hydraulic cement will cure in the presence of water.

Fill deep damage only to within half an inch of the surface. After the cement hardens, fill completely with a second layer. Trim and smooth all repairs while the cement is still workable.

Damp or wet basements usually require troubleshooting to determine the cause of cracks and holes. This may be either water seeping into the basement from outdoors or moisture from indoor air condensing on cool basement surfaces—or both. To find out which problem you're dealing with, dry an area about twelve inches square with a hair dryer and cover the dry spot with clear plastic sheeting. Use four- or six-mil polyethylene, not plastic kitchen wrap. Seal all four edges to the concrete with duct tape. Leave the patch in place until you see signs of moisture, usually within five days.

If the exposed surface of the plastic becomes moist, indoor condensation caused by excess humidity is at least part of the problem. Generally, this can be cured by the methods described throughout this chapter and in the chapter on radon for increasing ventilation and sealing basement surfaces. However, unless basement ventilation is quite good, excess humidity can develop in spite of these measures, especially during summer months. Then it may be feasible to close basement windows and use a dehumidifier to dry the air. If radon is a problem, installing a subslab depressurization system may be the only way to eliminate it and also keep the basement free of excess humidity.

If moisture accumulates behind the patch, it must be coming through the masonry from the ground. (There is one exception: in a brand-new house, concrete and other building materials may take six months to two years to lose excess moisture by evaporation.) Barring entry through visible leaks like those described above, the chief cause of moisture penetration through basement materials is hydrostatic pressure. This is the pressure exerted by saturated soil; it can be great enough to force water through the pores of solid masonry.

The first line of defense in combating hydrostatic pressure is minimizing moisture in the soil. Ironically, faulty gutter systems often are to blame. Make sure gutters work properly. Clean them of debris; if water collects along their length, check for dips caused by loose sections or supports. Caulk dripping seams. If gutters overflow, consider adding more downspouts or larger gutters (these often are necessary to drain large roofs; builders sometimes cut corners by installing gutters that are undersized). Also make sure gutters are correctly

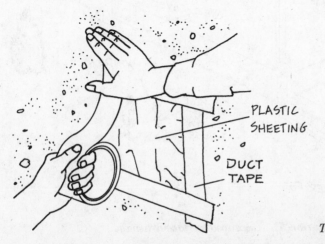

PLASTIC
SHEETING

DUCT
TAPE

Illustration 50.
Testing for condensation

pitched; they should slope noticeably toward downspouts when viewed against roof edges.

A downspout always must empty onto a splash block of concrete or other durable material that channels water away from the foundation. Even so, it may help to extend downspout drainage farther from the house by adding four-foot sections of straight downspouting to their elbows. If this is unsightly or interferes with mowing, consider lengthening the vertical run of the downspout and then burying the extension. The extension should run either to a "bubbler pot" (a partly buried container that allows accumulated water to flow out of

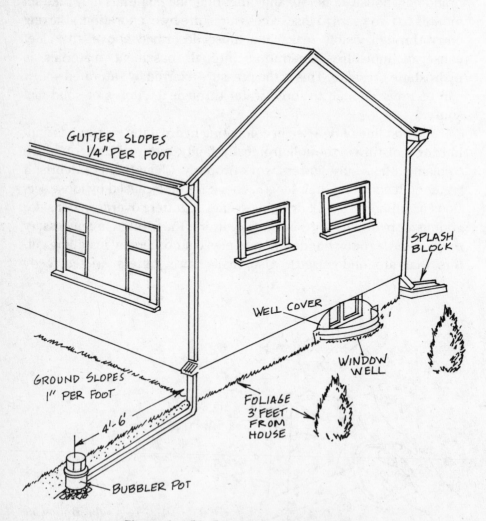

GUTTER SLOPES
1/4" PER FOOT

SPLASH
BLOCK

WELL COVER

WINDOW
WELL

GROUND SLOPES
1" PER FOOT

FOLIAGE
3' FEET
FROM
HOUSE

4'-6'

BUBBLER POT

Illustration 51. Improved surface drainage

it into the ground) or to a conventional dry well. The latter is a buried, porous container surrounded by and filled with coarse gravel. Water directed into it drains away rapidly underground.

Downspouts on older houses in urban areas often drain into underground pipes that lead to storm sewers. Over the years, these pipes break, become separated, or clog. Call your city's department of public works to find out whether servicing is available for such drain systems; if not, call a sewer-cleaning company to have the pipes inspected and cleared with a power auger. The cost at this writing averages about $150 for a house with two downspouts. Sewer-cleaning companies are listed in the telephone book; do not confuse them with septic tank cleaners.

Also check that basement window wells drain rapidly or are covered with plastic window bubbles to keep out rain and debris. Open window wells are often filled with leaves; clearing them should reveal a deep layer of gravel and, in many instances, the screened end of a vertical drainpipe leading to the horizontal system of foundation drain tiles buried at the level of the footing.

Trim back moisture-trapping foliage and shrubbery and eliminate deep flower beds to leave at least three feet of clear space around the house. Examine the slope of the ground near the foundation closely; areas that slope toward the house may need regrading so they slope away from it. Naturally, the method for regrading depends on individual circumstances, and you may want to call in a landscaping contractor to perform the task in any case. (Get references from clients of the contractor who have had regrading done; the job requires specialized, high-level skills and experience.) The object is to slope the earth away from the foundation at the rate of at least one inch per foot. After regrading, plant grass—whose root systems provide superior moisture absorption—instead of flowers or other ground cover.

If the conditions described above seem satisfactory and only basement walls are damp, coating them with a sealing compound formulated for the purpose might work. There are drawbacks to using sealers applied to the inside of basement walls: water pressure from outside can force the sealer away from the surface of the wall, and walls require exacting preparation before sealing that is often impossible to achieve.

Two basic varieties of sealer for basement walls are available; epoxy, which, like the patching material mentioned earlier, should be applied by professionals after an independent inspector's recommendation; and cementitious waterproofing paint. The latter contains fine cement particles (hence its name) that fill the pores of masonry, making entry by water from outside more difficult. The easiest to use and usually most effective type is premixed and contains acrylic resin.

Cementitious paint costs about the same as quality interior house paint (about $20 to $40 per gallon) but has a lower spread rate, meaning that a gallon does not cover as much area. Application, too, is similar to interior painting—except that a brush rather than a roller is best for applying the initial coat (two coats usually are required).

Do not rely on cementitious paint (or epoxy, for that matter) to work very well on walls that bear traces of previous painting unless the previous paint also was cementitious and is still in good condition. Oil, alkyd, and latex paints all prevent waterproofing products from entering and sealing masonry pores; also, these paints may eventually fail, loosening the bond of the waterproofing products applied over them.

To apply cementitious paint, start by repairing all visible cracks and holes, as was described earlier. Allow any patching products to cure, then clean the walls by scrubbing with a wire brush to remove loose particles. Finally, wash the walls with a mild acid to neutralize and remove efflorescence (a white, powdery residue caused by moisture).

Cleaning basement walls is messy and dusty. Beforehand, seal openings leading from the basement to the rest of the house with plastic and cover anything in the basement you do not want coated with dust. When working, wear safety goggles, heavy work gloves, and most important, a respirator or dust mask rated for protection against silica particles. The rating should be clearly printed on the package; do not use an inexpensive "comfort mask" as a substitute.

Brush walls by starting at the bottom and working upward; that way, you will be exposed to less dust. Vacuuming the floor and wall frequently with a shop vacuum helps also; in any case, vacuum the wall after you have finished and the dust has been allowed to settle for several hours.

To wash the walls, choose a product such as sulfamic or muriatic acid, both designed for removing efflorescence from masonry. Be advised that fumes from sulfamic acid are less noxious than those from

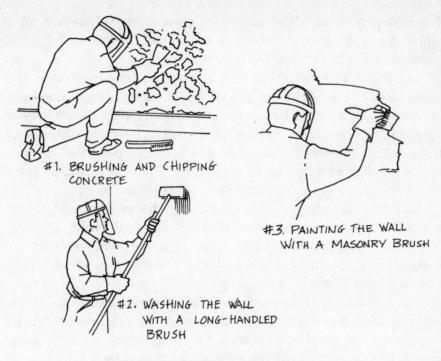

Illustration 52. Applying cementitious paint

muriatic acid. Whichever product you use, follow the mixing or diluting instructions and directions printed on the container.

Prepare the basement for washing by opening all windows and setting up fans to eliminate the acid fumes (exhaust fans can be rented). You should also make some provision for containing and disposing of the rinse water. Because spraying the walls with a garden hose is usually the best way to rinse them, one way to contain the water is to lay sand bags in front of the walls to form low dikes that will collect the water in pools for bailing or will channel it to the sump drain, at the basement's lowest point.

Use a long-handled bristle brush to scrub the walls with acid solution, and wear goggles or a full-face mask, long rubber gloves, and clothing that covers as much skin as possible to protect against splashes.

On application, the acid will bubble as it cleans the surface. When the bubbling stops (after about fifteen minutes), rinse the walls with plenty of water—using a hose, as was mentioned earlier, or a large, well-saturated sponge. The acid in the rinse water is harmless after bubbling has stopped. Allow the walls to dry overnight or longer; constant ventilation speeds the process.

The paint can now be applied; however, if temperatures in the basement are very low, make provisions for maintaining surface and air temperatures inside the basement of 50 degrees Fahrenheit or higher during application and curing. Spread on the paint with a large nylon or tampico-bristle paintbrush and force the paint into the surface of the wall with the tips of the bristles. Obey the manufacturer's guidelines for coverage to avoid spreading the paint too thinly.

Allow the first coat to dry for twenty-four hours, then apply a second coat the same way. Allow several days for final curing, and provide plenty of ventilation to allow moisture in the paint and in the air to escape. Occasionally, a third coat is necessary on areas where pores are large or surfaces are rough.

When coating walls is not likely to be effective (usually due to extreme hydrostatic pressure), or when the problem is moisture rising through the basement floor, probably the only solution will be some form of drainage that involves excavation. Again, get a professional inspection before proceeding.

Installing a sump pump can help. This requires breaking through the basement floor and digging a pit into which is inserted a perforated container—called a sump liner—about the size of a five-gallon bucket. Into the liner is placed a submersible pump with a drain hose running to the outside. The pump is electrically powered and is triggered by a switch that detects water rising in the sump liner.

Sump pump systems must have tight-fitting covers to prevent moisture and, in some cases, radon from rising into the basement. Drainpipes for sump pumps must extend from the pump to the outside and empty well away from the foundation, otherwise water flowing from them simply sinks back into the ground and travels back into the sump liner, where it is pumped out again, over and over. Submersible sump pumps cost around $100; having a sump pit dug and a pump professionally installed can cost $300 to $500, depending on difficulty.

A more expensive but highly effective drainage system is an interior footing drain. This is a length of perforated drainpipe installed around the perimeter of the foundation beneath the basement floor.

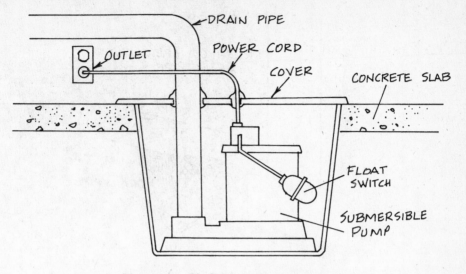

Illustration 53. Submersible sump pump

Excavating to lay the pipe involves breaking through the slab to create a trench along the base of each basement wall and may involve other digging to connect the pipe to a sump pump system or to a dry well located away from the house. Connections can also be made that enable the pipe to function as part of a subslab depressurization system for radon removal (see the chapter "Radon"). After the pipe is in place, gravel is added to the trench to enhance drainage and protect the pipe from clogging, then fresh concrete is poured to repair the slab.

An interior footing drain can remove large amounts of water; more important, it also relieves hydrostatic pressure beneath the basement floor. Costs for the project, as of this writing, average about $40 per foot of trenching required.

The most elaborate drainage method, often costing $100 per foot or more for an average house, is excavating to expose the entire foundation and then installing a waterproofing membrane against the outside of the foundation walls, followed by a footing drain and downspout collection system, both with clean-out traps for periodic maintenance.

Naturally, such a system is best installed during new construction. Several types of waterproofing membrane exist. For remodeling, the best probably is rigid foam drainage panel; installing it

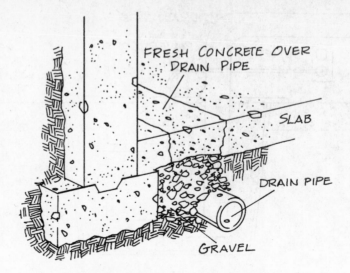

Illustration 54. Interior footing drain

requires less rigorous cleaning of the foundation walls than does installing spray-on or adhesive types consisting largely of asphalt.

The footing drains, which consist of perforated pipe like that described above, must be surrounded by gravel; and indeed, the entire trench should be filled with gravel to a depth of sixteen to twenty inches below ground level. Filter fabric, a spun synthetic that resists penetration by silt and other particles but readily admits water, should be placed against the outside of the trench as it is filled and then laid over the top of the gravel and against the foundation to cover it before the final soil is added. The fabric prevents soil from mixing with the gravel, which would reduce the ability of water to flow downward to the footing drain and diminish the capacity of the loose stones to dissipate hydrostatic pressure.

The downspout collection system can simply be as was described earlier—individual downspout extensions running to buried dry wells—or it can be a network of pipes linking all the downspouts and carrying away runoff through a single outlet.

Humidifiers and Dehumidifiers

While building codes require vapor barriers in virtually all new construction, they are—as was mentioned earlier—difficult to retro-

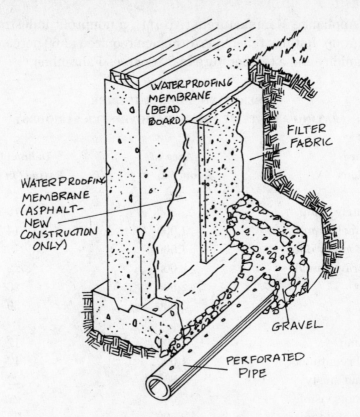

WATERPROOFING
MEMBRANE
(BEAD
BOARD)

FILTER
FABRIC

WATERPROOFING
MEMBRANE
(ASPHALT-
NEW
CONSTRUCTION
ONLY)

GRAVEL

PERFORATED
PIPE

Illustration 55. State-of-the-art exterior foundation drainage

fit in existing homes. Too, they may not be effective against excess dryness. The solution may be a humidifier or a dehumidifier, or both. Humidifiers add moisture to air; dehumidifiers remove it. Both appliances are available as portable models; humidifiers also are available as permanent units that can be added to forced-air central heating and cooling systems.

Consider a dehumidifier for drying specific areas in a house or for use during especially humid weather. The cost of these appliances is comparable to that of air conditioners; both have similar parts and function in nearly the same way.

When selecting a dehumidifier, first consider the conditions under which it will be used and the space it must service. Ratings of models usually are in pints of water an appliance can remove during a twenty-four-hour period in a room having a given area. The following chart is based on figures and descriptions approved by the Association of

Home Appliance Manufacturers (AHAM), a nonprofit industry-regulating group. Ratings reflect performance required at 60 percent relative humidity and a temperature of 80 degrees Fahrenheit.

SELECTING A DEHUMIDIFIER
(RATING = PINTS OF WATER REMOVED PER 24 HOURS)

Moisture Conditions	Area of Room (sq. ft.)	Dehumidifier Rating (in sq. ft.)
Moderately damp	500	10
(space feels damp	1,000	14
and has musty odor	1,500	18
only during humid	2,000	22
weather)	2,500	26
	3,000	30
Very damp	500	12
(space is continually	1,000	17
damp and musty;	1,500	22
damp spots visible	2,000	27
on walls or floors)	2,500	32
	3,000	37
Wet		
(space feels and	500	14
smells wet; walls	1,000	20
or floors show	1,500	26
moisture or seepage)	2,000	32
	2,500	38
	3,000	44

Also check the recommended low temperature for operating the appliance. Most dehumidifiers for residential use will develop frost when operated in rooms where the temperature is below 65 degrees Fahrenheit. Frost can damage a dehumdifier's fragile evaporator coil, which contains refrigerant. Better dehumidifiers are equipped with a frost control that shuts down the appliance when frost develops and then turns it back on when the frost has melted. The best models actually defrost the coil. Any dehumidifier you buy should be listed by Underwriters Laboratories and carry the "UL" symbol. There should be an automatic humidistat control that turns the appliance on and off to maintain a selected humidity, and an automatic shutoff control to prevent overflowing. Make sure also that the moisture container is large but manageable, has a light or alarm that indicates when it is full, and is easily removed for emptying.

Consider a humidifier if indoor air is too dry. Small, so-called tabletop models that deliver enough moisture to humidify a small room such as a bedroom range in price from about $20 to about $75. Larger console models on wheels resemble dehumidifiers in size and have the capacity to humidify areas up to around 3,000 square feet. Their cost ranges from about $75 to about $175.

Small console humidifiers deliver about eight gallons per twenty-

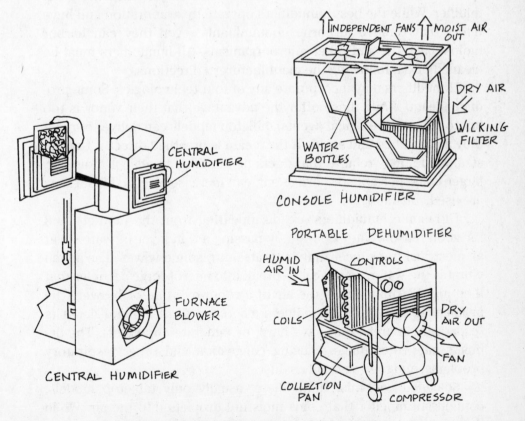

Illustration 56. Humidifiers and dehumidifiers

four-hour period—twice the amount of a large tabletop model—while the largest consoles can put out as much as fifteen gallons in twenty-four hours.

Portable humidifiers are versatile and relatively inexpensive, but a drawback is that water must be added continually. This can be a chore with both very small units that must be frequently refilled and large consoles that hold many gallons. The answer can be a permanent, central humidifier attached to a forced-air heating system. These units usually cost less than $200 (professional installation, which is recommended, can add another $150) and are directly connected to a home's main water supply. Central humidifiers operate only when the heating or cooling system is running, but in most houses the system is to blame for drying the air in the first place.

If anyone in the household has allergies, and especially asthma, consult an allergist or respiratory specialist before installing a humidifier. While the best humidifiers operate by evaporation and have filters that remove airborne contaminants, even they can harbor molds, bacteria, and other microorganisms. All humidifiers must be cleaned often, following the manufacturer's directions.

Humidifiers may incorporate any of four technologies. Some produce steam. While these offer the advantage that their vapor is too hot to support biological agents, tabletop models can deliver burns if they are touched and scalds if the steam is breathed directly. Central steam units may require that ducts be modified to accept moist air; larger units, portable or permanent, can increase electric bills if they are used constantly.

Ultrasonic humidifiers are disappearing from the market—and for good reason. They operate by passing a thin layer of water over an electrical transducer that generates ultrasound waves. The sound vibrates the water, causing it to break into microscopic particles that then are distributed into the air of a room by a fan or blower. The problem is that these humidifiers also send forth mineral deposits (especially in areas that have hard, or mineral-rich, water). The deposits can enter lungs, causing congestion and other respiratory problems, and they also fall as dust.

So-called cool-mist humidifiers, usually only tabletop models, contain an impeller that spins moisture droplets into the air. While they do not create mineral dust, as do ultrasonic humidifiers, they can distribute microorganisms. These small units use electricity only for the impeller, not for heating water. Therefore, they have a smaller effect on energy bills than do steam humidifiers.

Evaporative humidifiers, as was mentioned earlier, generally are considered the best. These consist of a wet medium—a saturated pad or filter—through which air is blown. The air wicks away evaporating moisture, and this moisture-laden air is blown through ducts or directly into a room. Solids, such as biological contaminants and mineral deposits, remain behind. Evaporative tabletop and console humidifiers contain their own fans; a permanent unit usually relies on the blower of the central heating or cooling system to which it is attached, which accounts for the relatively low cost of these units.

More than with dehumidifiers, it is important to select a humidifier that suits the size of the area it is intended to serve. If a humidifier is too small it will be ineffective and simply a bother and energy-waster. A too-large humidifier can be a disaster. Gather advice from a licensed heating contractor or engineer before shopping for a humidifier, and buy only a unit certified by the AHAM and listed by Underwriters Laboratories. Evaporative capacities specified should be in accordance with Air-Conditioning and Refrigeration Institute Standard 610-82.

SELECTING A HUMIDIFIER
(CAPACITY = GALLONS OF WATER INTRODUCED PER 24 HOURS)

Type of House Construction	Area to Be Humidified (sq. ft.)	Humidifier Capacity
Loose (home has	500	5
little insulation,	1,000	9
vapor protection	1,500	16
or weatherstripping,	2,000	18
no storm doors or	2,500	25
windows)	3,000	NA
	3,500	NA
Average (home is	500	3
insulated, features	1,000	6

(continued)

SELECTING A HUMIDIFIER
(CAPACITY = GALLONS OF WATER INTRODUCED PER 24 HOURS) *(cont'd)*

Type of House Construction	Area to Be Humidified (sq. ft.)	Humidifier Capacity
loose storm doors	1,500	9
and windows, dampered	2,000	14
fireplace, minimal	2,500	16
weatherstripping)	3,000	19
	3,500	24
Tight (home is well	500	2
insulated and sealed	1,000	4.5
with tight storm doors	1,500	6.5
and windows, vapor	2,000	8.5
barrier, thorough	2,500	10.5
weatherstripping, other	3,000	13
"tight-house" features)	3,500	15

Controls for portable humidifiers should include a humidistat, fan speed and airflow regulators, and a low-water warning light. Some models also have a light indicating when the filter is dirty and needs replacing. Look for low-noise units.

On permanent humidifiers, look for easy cleaning and maintenance features (for example, a see-through window and a large removable panel for accessing the wet medium), stainless steel and plastic parts that resist corrosion, automatic shutoff and water-level controls, and (on powered models) a self-reversing motor to prevent burnout. Special humidifier designs are available for use with hard and extremely hard water.

Air Cleaners

Mechanical and electronic air cleaners have become popular, and most do a reasonable job of removing fine dust and combustion particles, smoke, and some biological contaminants from indoor air. But air cleaners will not remove gases such as radon and carbon monoxide, and some odors, notably tobacco; nor will they remove extremely small organisms such as viruses—or relatively large particles such as pollen unless they are suspended in the air, which seldom is the case. Most allergy-causing particles—pollen, dust mite residue, and animal dander—are large enough to settle out of the air and onto surfaces fairly quickly; as a result, even though an air cleaner can remove such particles from air, installing one for the purposes of allergy relief often fails to achieve the desired result because so many allergy-producing particles remain. However, when they are coupled with sufficient household ventilation and reduced humidity, air cleaners can benefit indoor air quality.

Tabletop, console, and permanent varieties of both mechanical and electronic air cleaners are available. Prices range from about $50 for a tabletop model to around $1,000 for a top-of-the-line permanent model professionally installed as part of a forced-air central heating or cooling system. Few, if any, tabletop air cleaners have been found effective except in very confined spaces like office cubicles. Consoles, or other portable models, are best for cleaning air in bedrooms and other individual areas of a house; expect to pay $150 to $350 for a worthy console air cleaner.

Mechanical air cleaners draw air by means of a fan through a series of filters—the best draw the air through a foam prefilter that traps large particles that could cause clogging and then through a HEPA filter that can trap virtually the smallest particles, including those that make up smoke. Most models also contain an activated carbon or charcoal filter that removes many types of odor-causing molecules, and some models contain a postfilter for arresting remaining particles that the fan blows loose from the other filters. Fans may be propeller (axial) or squirrel cage (centrifugal) type, depending on the manufacturer.

Other mechanical air cleaners contain a media filter, which is made of dense, ultrafine fibers and is usually pleated to provide maximum surface area. Media filters can also be installed on forced-air central heating and cooling systems at a cost of around $300. Professional sizing and installation are necessary because the filters alter air pressure within the system; this sometimes requires adjusting or modifying other components.

Electronic air cleaners fall into three categories: electrostatic, electret, and negative-ionizing. In an electrostatic air cleaner, parti-

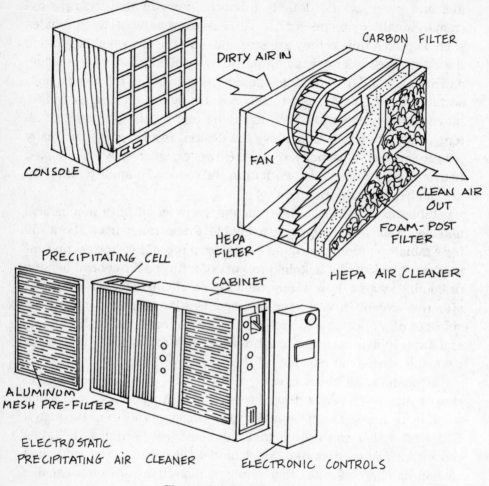

CONSOLE

DIRTY AIR IN

CARBON FILTER

FAN

CLEAN AIR OUT

FOAM-POST FILTER

HEPA FILTER

HEPA AIR CLEANER

PRECIPITATING CELL

CABINET

ALUMINUM MESH PRE-FILTER

ELECTROSTATIC PRECIPITATING AIR CLEANER

ELECTRONIC CONTROLS

Illustration 57. Air cleaners

cles drawn by the fan into a chamber called a precipitating cell are ionized (charged) with positive electricity from a grid of high-voltage wires and then are attracted onto negatively charged aluminum collectors as the air leaves the cell. In an electret air cleaner, a special filter—the electret—is charged with static electricity; the charge attracts particles entering the cleaner, and the particles cling to the filter.

Self-charging electret filters have been developed for use instead of conventional fiberglass furnace filters in forced-air heating systems. These are sometimes called electrostatic filters. Their ability to

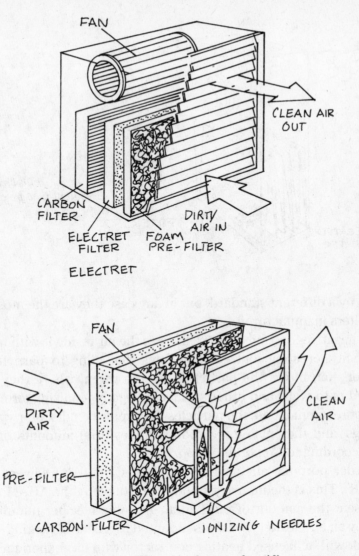

FAN

CLEAN AIR
OUT

CARBON
FILTER

ELECTRET
FILTER

FOAM
PRE-FILTER

DIRTY
AIR IN

ELECTRET

FAN

DIRTY
AIR

CLEAN
AIR

PRE-FILTER

CARBON FILTER

IONIZING NEEDLES

Illustration 57A. Air cleaners (cont'd)

trap particles is significantly greater than that of ordinary fiberglass filters, but neither type really has much effect. By the industry-standard testing methods developed by the American Society of Heating, Refrigeration, and Air Conditioning Engineers (ASHRAE), ordinary fiberglass furnace filters are about 5 percent efficient; electrostatic filters are only about 20 percent efficient. Media filters test at between 35 percent and 88 percent efficient. HEPA filters are mea-

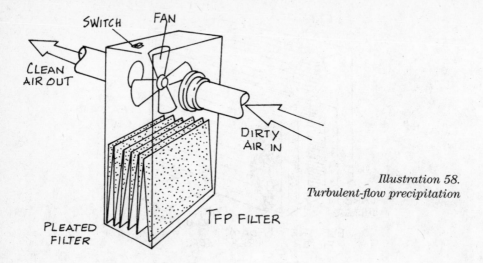

Illustration 58.
Turbulent-flow precipitation

sured by a different standard, but in any case they are the most effective filters manufactured.

A negative-ionizing cleaner charges the air inside it with negative electricity, creating ionized molecules that cling to particles. The cleaner's fan blows the particles into the room, where they are attracted onto walls and other surfaces. Negative-ionizing air cleaners are not recommended; the particles they spew can discolor walls and ceilings, and they also can produce unhealthful amounts of ozone gas, according to *Consumer Reports*.

Order portable air cleaners by their Clean Air Delivery Rating (CADR). This is the industry standard certified by the AHAM; ratings represent the amount of purified air in cubic feet per minute delivered by an air cleaner. For units mounted to a central heating system, consult with a licensed heating contractor who has experience in installing such equipment.

Air cleaners can have the same CADR ratings but different rates of airflow, meaning they process air at different speeds. Choose the cleaner with the highest airflow rate. Air cleaners with HEPA filters do a better job than other types at removing particles; however, electrostatic cleaners generally have higher airflow rates. Therefore, an electrostatic cleaner with a lower CADR rating may be able to purify air as effectively as a cleaner with a HEPA filter and a higher CADR because the former can process the same air several times—cleaning

SELECTING AN AIR CLEANER

Room Area (sq. ft.)	Air Cleaner CADR
80	50
120	80
144	100
192	120
224	145
320	200
360	230
400	250
480	300

it more with each pass—in the interval it takes the latter to process the air once.

A new type of air cleaner, called a turbulent flow precipitator, recently has become available. This type might be considered a hybrid mechanical air cleaner. While it has no filters as such, neither is it electrostatically charged as are electronic air cleaners.

A turbulent flow precipitator consists of two air spaces enclosed in a rectangular metal box, which is attached to central heating or cooling ductwork like the other varieties. Particle-laden air flows through the upper air space, which is continuous with the ductwork, and passes over the top edges of filter sheets arranged vertically in the lower space. The filter sheets are attached to each other accordion-style; the spaces between the sheets contain dead air into which particles fall from the air passing above.

Surprisingly, the system so far has demonstrated remarkable filtering results, according to the manufacturer, who claims nearly 100 percent removal of coarse particles like pollen, fungus, dust, and dirt, and 90 percent removal of fine particles including tobacco smoke, animal dander, skin flakes, and some bacteria. Maintenance consists of yearly cleaning, and replacement of filters (which are inexpensive) about every four years.

Volatile Organic Compounds

What Are They?

VOLATILE ORGANIC COMPOUNDS (VOCs) ARE CARBON-BASED chemical compounds that exist as vapors—gas—at room temperature. Unlike the substances covered in previous chapters, VOCs do not occur naturally; they are by-products of substances created by the chemical industry, and also result more or less unintentionally from other manufacturing processes and from combustion. Formaldehyde is one of the most common and thoroughly analyzed VOCs. Others include compounds with familiar names such as benzene, vinyl chloride, toluene, xylene, styrene, methyl chloride, and acetone; there are many more whose names probably are less familiar, such as para-dichlorobenzene, 1-1-1 trichloroethane, trichloroethylene, pentachlorophenol, 4-phenylcyclohexene, and polychlorinated biphenyls (also known as PCBs).

Materials and products containing VOCs release them in varying quantities by a process similar to evaporation called outgassing, in which unstable molecules escape into the atmosphere, especially under conditions of increased heat and humidity. Breathing or otherwise absorbing even small doses of some VOCs (PCBs, among others, can pass directly into the body through the skin) can cause a variety of unpleasant symptoms: headache; irritation of the eyes, nose, throat, and lungs; fatigue; drowsiness; dizziness; nausea; joint pains; chest tightness; tingling sensations in the extremities; blurred vision; euphoria; unsteadiness; and irregular heartbeat. Prolonged exposure to high doses of certain individual VOCs can have harmful, frequently irreversible results—including chronic upper respiratory

148

infections; liver and kidney damage caused by scarring; metabolic, menstrual, and fertility disorders; and cancer.

The effects of prolonged low-dose exposure are not fully known—indeed, individual tolerance of VOCs varies widely, with some people exhibiting no symptoms at dosages that cause severe discomfort to others—but since the late 1980s researchers have come to recognize a condition called solvent encephalopathy, characterized by a group of neurophysiologic symptoms occurring together (including headache, irritability, poor concentration, and coordination difficulties), that appears to be associated with low-level exposure to multiple VOCs. Symptoms of solvent encephalopathy seem to occur at dosage levels lower than those required to produce symptoms from exposure to individual VOCs; evidence also suggests that the duration of exposure influences how long recovery may take after exposure stops.

How Bad Is the Problem?

So far, over nine hundred VOCs, of which about fifty are common, have been identified in indoor air. According to the EPA, "these compounds are incorporated into almost all materials and products that are used in construction materials, consumer products, furnishings, pesticides, and fuels." That is to say, they are virtually everywhere.

Formaldehyde has been singled out among VOCs by researchers partly because of its long and widespread use. Discovered in 1867 by the German chemist August von Hofmann, formaldehyde was instrumental in the creation of the first entirely synthetic plastic material—phenol-formaldehyde, or Bakelite—developed in 1909 by the Belgian-American scientist Dr. Louis Baekland. Since then, formaldehyde has played a key role in the production of other synthetic resins that have been and continue to be crucial to many of the world's largest manufacturing industries.

Another, more significant reason for its special study is that formaldehyde is the chief ingredient of urea-formaldehyde, a controversial compound used during the 1970s as an insulating foam. Approximately 435,000 homes were insulated with this material between 1975 and 1981, and some 1,500 complaints from occupants were received by the Consumer Product Safety Commission (CPSC). These ranged from accounts of headaches and dizziness to those of nosebleeds, vomiting, and skin rashes. In 1979, the Chemical Industry Institute of Toxicology (CIIT) announced research results indicating that formaldehyde had produced nasal cancer in laboratory rats.

Both circumstances led the CPSC to ban urea-formaldehyde foam as residential insulation in August 1982. However, the ban was overturned nine months later on appeal by the Formaldehyde Institute, a trade association, which successfully disputed the claim that formaldehyde was a human carcinogen. (In the CIIT tests, 103 of 240 rats developed cancer after exposure to 14.3 parts formaldehyde per million parts of air. By contrast, formaldehyde levels in homes typically range from about .04 parts per million (ppm) to .5 ppm, with the average nowadays less than .1 ppm.)

Research on the problem continues, and it appears that urea-formaldehyde foam has little effect on human health if its components are mixed properly on-site under dry conditions that permit the foam to harden. (If it is too wet, urea-formaldehyde foam remains soft, which increases and prolongs outgassing.) Indeed, so far not one case of human cancer has been successfully attributed to residential formaldehyde exposure; but because several states have effected their own bans and because of widespread negative public opinion, the use of urea-formaldehyde foam has virtually ceased. Currently, the EPA, the Federal Panel on Formaldehyde, the International Agency for Research on Cancer, and other similar organizations classify formaldehyde as a "probable human carcinogen."

The greatest sources of formaldehyde in homes (apart from those few houses that contain urea-formaldehyde foam insulation) are manufactured wood products, especially medium-density fiberboard (MDF), particleboard used for floor underlayment, and hardwood plywood used for cabinets, furniture, and indoor paneling. Each of these contains urea-formaldehyde resins as the chief ingredient in the adhesive binding together its fibers, particles, or plies; on the other hand, the U.S. Department of Housing and Urban Development (HUD) set standards on these products in 1984, limiting formaldehyde emissions to .2 ppm for plywood and .3 ppm for the other wood products mentioned. Softwood plywood, for exterior use, is made

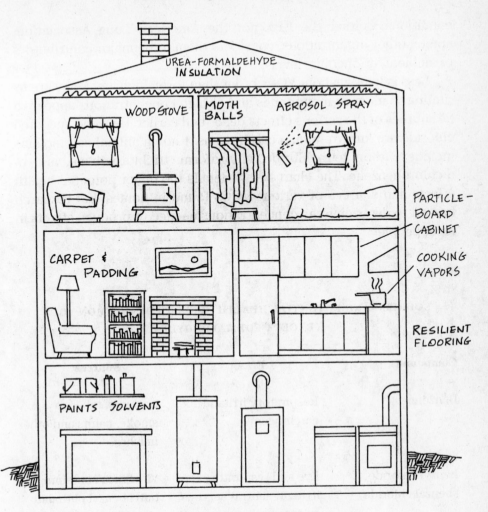

Illustration 59. Sources of VOCs in a house

with phenol-formaldehyde resins that are more resistant than urea-formaldehyde to moisture; phenol-formaldehyde products emit virtually no formaldehyde.

Other sources are combustion by-products from smoking; emissions from kerosene, wood, coal, and gas stoves (these are discussed in the chapter "Combustion Products"); and emissions from carpet and carpet padding, resilient flooring, gypsum wallboard, fiberglass insulation, ceiling tiles, draperies, upholstery materials, and clothing. However, it is important to note that while these sources have the ability to raise indoor formaldehyde levels significantly under certain conditions, neither the EPA nor the American Lung Association, among other organizations, considers them to be major contributors to ambient, or ongoing, levels.

Less is known about VOCs other than formaldehyde. Many are irritating to mucous membranes and at high concentrations appear to be sources of the serious effects described earlier. Benzene and vinyl chloride are known human carcinogens. Known animal carcinogens include carbon tetrachloride, chloroform, trichloroethane, and p-dichlorobenzene. The chart below details the major potential health effects and sources of various VOCs found in homes, but research has yet to determine the concentrations needed to produce effects in most people.

SOURCES AND POTENTIAL HEALTH EFFECTS OF COMMON VOCS (EXCEPT FORMALDEHYDE)

Compound	Effect	Sources
Benzene	Respiratory irritant, carcinogen	Solvents, tobacco smoke, paints and other finishes
Benzyl chloride (benzal chloride)	Eye and respiratory irritant, central nervous system depressant, causes kidney and liver damage	Vinyl tiles containing butyl benzyl phthalate plasticizer
2-butanone	Eye and respiratory irritant, central nervous system depressant	Fiberboard, particleboard, caulking compounds, tobacco smoke
Chlorobenzenes	Narcotic, causes kidney and liver damage, causes pulmonary damage	Paints and other finishes, solvents

Volatile Organic Compounds

Compound	Effect	Sources
Ethyl benzene	Eye and respiratory irritant; central nervous system depressant	Solvents
Isopropanol (rubbing alcohol)	Central nervous system depressant	Household cleaners
Methanol (wood alcohol)	Central nervous system damage; produces formaldehyde in the body when ingested, often leading to blindness and death	Paints, adhesives, wood finishes, solvents, dyes
Methylene chloride (dichloromethane)	Narcotic, central nervous system depressant, probable carcinogen	Paint remover
Methyl ethyl ketone (acetone)	Eye and respiratory irritant, central nervous system depressant	Paints, finishes, solvents, cleaning fluids, nail polish remover
Para-dichlorobenzene	Narcotic; eye and respiratory irritant; causes liver, kidney, and central nervous system damage	Moth repellent crystals, air fresheners
Petroleum distillates	Central nervous system depressant, causes liver and kidney damage	Solvents, cleaning products, roofing and asphalt supplies
4-phenylcyclohexene	Eye and respiratory irritant, causes central nervous system damage	Carpet adhesive

SOURCES AND POTENTIAL HEALTH EFFECTS OF COMMON VOCS
(EXCEPT FORMALDEHYDE) *(cont'd)*

Compound	Effect	Sources
Polychlorinated biphenyls (PCBs)	Probable carcinogens	Electrical components; some paper and plastic products
Styrene	Narcotic, eye and respiratory irritant, central nervous system depressant, possible carcinogen	Plastics
Toluene	Narcotic, possible cause of anemia	Solvents, adhesives, wallpaper, joint compound, vinyl flooring, caulking compounds, paint, chipboard, tobacco and kerosene smoke
Toluene diisocyanate (TDI)	Sensitizer (lowers dosages of other VOCs required to produce symptoms), probable carcinogen	Polyurethane foam aerosols
1,1,1-trichloroethane	Possible carcinogen	Solvents, aerosols
Trichloroethylene	Causes central nervous system damage, probable carcinogen	Solvents, dry-cleaning products
Xylene	Narcotic, eye and respiratory irritant, central nervous system depressant, causes liver and kidney damage, causes heart damage	Adhesives, joint compounds, floor coverings and finishes, solvents, dyes, tobacco and kerosene smoke

Volatile Organic Compounds

FORMALDEHYDE CONCENTRATIONS, STANDARDS, AND EFFECTS

Health effect	Standard	Concentration (parts per million)
None		0.0–0.5
Odor threshold		0.05–1.0
Neurophysiologic (headache, dizziness, related symptoms)		0.05–1.5
Eye irritation		0.01–2.0
Upper respiratory irritation		0.10–25
	Maximum level for continual exposure, recommended by the American Society of Heating, Refrigeration, and Air-Conditioning Engineers (ASHRAE)	0.10
	Limit set by U.S. Department of Housing and Urban Development (HUD) for formaldehyde emissions from interior plywood used in mobile homes and other manufactured housing	.20
	Concentration threshold below which 80 percent of the population is unlikely to experience symptoms, according to the National Academy of Sciences	.25

FORMALDEHYDE CONCENTRATIONS, STANDARDS, AND EFFECTS (*cont'd*)

Health effect	Standard	Concentration (parts per million)
	Limit set by HUD for formaldehyde emissions from particleboard used in mobile homes and other manufactured housing	.30
	Overall formaldehyde emissions standard suggested by HUD for mobile homes and other manufactured housing	.40
	Standard set by the United States Occupational Safety and Health Administration (OSHA) for exposure in workplaces per 8-hour period	1.0
	Standard set by OSHA for exposure in workplaces not to exceed 15 minutes	2.0
Lower respiratory and pulmonary effects		5.0–30.0
Pulmonary edema, inflammation, pneumonia		50.0–100.0
Death		<100.0

Testing

Testing for VOCs is impractical unless someone in a household is suffering symptoms. Even then, homeowners usually are better advised to obtain professional testing and analysis than to attempt the task on their own; the reason behind this is that VOC emissions generally are of numerous types, each of which requires a separate test. For advice call the local health department, listed in the telephone book.

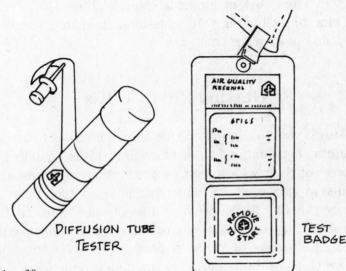

DIFFUSION TUBE TESTER

TEST BADGE

Illustration 60.
Formaldehyde testers

Kits for homeowners are available to test specifically for formaldehyde. The most accurate of these kits, according to some authorities, are the so-called DNPH devices, which are absorbent badges containing the chemical compound 2,4-dinitrophenyl-hydrazine. One of these devices is suspended from a ceiling so it hangs at average breathing height. Exposure should last from fifteen minutes to four hours, depending on conditions specified in the directions; afterward, the device is carefully repackaged and sent to a laboratory for analysis.

Unfortunately, DNPH devices are difficult to obtain singly. The reason is pure economics: because the homeowner market is small, manufacturers package the devices only in large quantities (ten or more) and target sales toward professional testers and hygienists. Currently, the cost for a package of ten DNPH testing devices is about $120. Laboratory analysis raises the price by about $40 per device.

Less accurate but more easily obtained are diffusion tube testers. These consist of a glass vial containing filter paper saturated with sodium bisulfate. Testing involves suspending the vial from a ceiling for five to seven days, then returning the device to a laboratory. Home centers and some large pharmacies carry diffusion tube formaldehyde testers. Their cost, including lab fees, is about $40.

Test kits for evaluating VOC emissions other than formaldehyde currently do not exist for amateurs.

Getting Rid of VOCs

The surest way to eliminate VOCs from a house is to avoid materials or products containing them. Of course, this is virtually impossible because of their widespread use; nevertheless, remodelers can take pains to use plywood and particleboard that bear grading stamps assuring that formaldehyde emissions are below HUD limits; and they should use only those paints, solvents, and finishes that contain reduced amounts of VOCs or none at all. (VOC-free or VOC-reduced versions of these and other building products are gradually becoming available through selected building supply stores and directly from manufacturers. It is likely that their number and distribution will dramatically increase. So, before starting a remodeling project, thoroughly investigate the scope of materials currently available.)

With formaldehyde, at least, time can remedy high concentrations. New products release much of any formaldehyde they contain within the first six months after being installed, and after a year levels can be expected to have dropped by half. Remaining levels generally drop by half again in less than four years, although with products manufactured during the last decade (these typically contain less formaldehyde to begin with) the second phase of reduction may take seven to eight years. In all cases, outgassing continues until no more unstable formaldehyde forms or is present.

Sealing materials containing formaldehyde is also effective, but care must be taken not to introduce other, possibly more harmful or long-lasting VOCs into the house by using a sealer containing them. New kitchen cabinets and other built-ins usually emit the largest amounts of formaldehyde because they often are made entirely of fiberboard or particleboard that can release the gas from unfinished edges or other surfaces. Coating unfinished areas with water-based polyurethane or lacquer is the best way to seal them; ordinary latex paint is ineffective.

Unfortunately, the factory finish on some cabinets contains formaldehyde that cannot be sealed except by refacing. One way to deal with this problem when buying new cabinets is to unpack and store them for two to six weeks in a garage or other well-ventilated area, preferably one in which temperatures can be significantly increased. During this time, especially when temperatures are elevated, the greatest degree of outgassing will take place; thus, less formaldehyde will escape from the cabinets after they are brought inside and installed.

Increasing heat and ventilation can also be used to speed up outgassing from freshly installed adhesive-backed carpet, tiles, resilient flooring, and other products such as caulking compound and solvent-based floor finish. However, because VOC levels are likely to become quite high during the process, this method is feasible only if the area being treated can be closed off from the rest of the house and remain unoccupied for the duration. Knowledgeable flooring contractors usually recommend staying away from a house for twenty-four to forty-eight hours after adhesive-backed carpeting or solvent-based floor finish has been applied, and they advise maintaining increased ventilation—although with normal heating and humidification—for several weeks or months afterward. When buying products such as carpet adhesive and caulking compound, it goes without saying that you should look for water-based varieties containing minimal VOCs. If such formulations are not available, select products with the fastest drying times (these release the largest portion of their VOCs in the shortest period of time). According to EPA research, fast-drying compounds may take only ten days to stabilize; slow-drying compounds, on the other hand, may release significant contamination for up to a year.

Upgrading ventilation and lowering indoor humidity levels can help cure a house of many indoor air quality woes. For a full discussion of the first topic, see the chapter "Improving Ventilation." The latter topic is covered in the chapter "Biological Contaminants."

Improving Ventilation

MOST HOUSES RELY ON NATURAL VENTILATION—BREEZES AND convection currents caused by warm air rising and cool air falling—to remove stale air and replace it with a fresh supply. But in many homes, particularly those that are well insulated and tightly sealed to reduce drafts, natural ventilation can be inadequate some or all of the time.

Occupants of a house need about fifteen cubic feet of fresh air per minute (cfm) for healthy, comfortable living, according to tests performed by the American Society of Heating, Refrigeration, and Air-Conditioning Engineers (ASHRAE), who established the figure in 1989. ASHRAE and other industry groups, including the Home Ventilating Institute (HVI), have determined that living spaces generally will supply this amount of fresh air provided the air in the space is exchanged with fresh air at a rate of at least 0.35 "air changes per hour" (ach). Note that in reality one air change does not remove all of the air in the space; because of mixing during the exchange, only about half the air is new after one air change.

ASHRAE figures take into account breathing at normal levels and the need for removing by-products of living, such as moisture and carbon dioxide, but they do not accommodate removing indoor pollutants such as formaldehyde and other VOCs, smoke from cooking, large amounts of carbon dioxide from exercising, or moisture from laundry, showers, and basement dampness. For this reason it is wise to consider the ASHRAE standards as minimal limits and to exceed them whenever possible, even at the expense of raising energy bills (by a manageable extent, of course).

There are strategies for improving ventilation that are more effective and less expensive than others. Probably the best overall is a combination of intermittent (or spot) ventilators for eliminating occasional pollutants such as bathroom moisture and cooking smoke, coupled with a continuously operating low-speed fan or other device that maintains an overall healthy ventilation rate.

The improvements mentioned in this chapter contribute to this strategy. In most houses they can be performed without undue consideration of possible effects other than those specific to their intended purpose, but if your house is extremely well insulated and weathertight you should take into account their possible effect on indoor air pressure. Imbalanced ventilation, one cause of which can be adding too many exhaust appliances, can produce a condition called backdrafting. When this occurs, potentially hazardous compounds released into chimneys and other vents can be drawn backward into the house.

Backdrafts

Backdrafts are the result of depressurization. When air inside a house is expelled faster than new air from outside can enter to replace it, the interior of the house contains less air—and, therefore, less air pressure—than is normal and is said to be depressurized. Because of the laws of physics, as the difference between indoor and outdoor pressures grows, so too grows the force of the outdoor air seeking to infiltrate the house and equalize the imbalance. At a certain point, the force of the outdoor air becomes sufficient to hinder the natural ability of warm air from indoor combustion sources to rise—it can even counter the ability of exhaust fans to move such air mechanically. When that happens, normal airflow within a house and through vent openings can be halted or reversed. The slowing or halting of exhaust air produces a condition called spillage. Reversal of the airflow is called a backdraft.

To roughly test for backdrafting, shut off your furnace and let it cool for at least fifteen minutes. Close all windows, exterior doors, and the fireplace damper (if there is one). Then turn on all household exhaust equipment that might operate at one time, including bath-

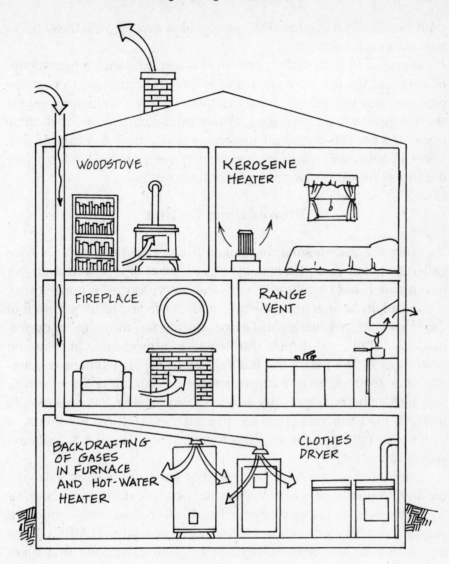

WOODSTOVE

KEROSENE HEATER

FIREPLACE

RANGE VENT

BACKDRAFTING OF GASES IN FURNACE AND HOT-WATER HEATER

CLOTHES DRYER

Illustration 61. Backdrafting

room fans, range hood, clothes dryer, and whole-house fan or other ventilation unit.

Restart the furnace and, if your water is heated by gas, turn on the hot water and let it run until the water heater goes on. Light a stick of smoky incense and hold it near the furnace and water heater draft hoods and dampers. If the smoke blows away from these openings, suspect backdrafting.

If you have a fireplace, perform this test also with a healthy fire blazing. Let the fire burn strongly for half an hour or so to heat the chimney and encourage a strong upward flow of air, then repeat the incense test as described. Safety equipment stores and some home centers sell backdraft test strips that attach to furnaces, gas water heaters, and other appliances. On one popular kind, a white dot turns black if prolonged backdrafting occurs.

Blower Door Testing

More accurate testing for backdrafting is possible with a device called a blower door. Essentially, this consists of a powerful fan set into a panel that temporarily seals the entry door of a house. After closing all other openings, the fan can be turned on to blow air out of the house, depressurizing it. Gauges on the fan measure its performance (airflow) and display differences in atmospheric pressure on both sides of the panel; the building's relative airtightness is determined by the amount of airflow required to maintain a depressurization inside the house of fifty pascals (abbreviated Pa); pascals are units of measure reflecting the pressure exerted by a column of water. The tighter the house, the less airflow is needed to maintain depressurization.

Additional calculations derived from airtightness figures can determine the point at which spillage and backdrafting will occur, or whether either is presently occurring. Blower door tests conducted for the Canadian Mortgage and Housing Corporation (CMHC) found that masonry fireplaces showed signs of backdrafting in houses with negative indoor pressures of three pascals; natural-draft furnaces and water heaters showed similar signs at negative indoor pressures above five pascals. (Depressurization can also increase radon levels in some houses. For more on this, see the chapter "Radon.")

Because the technology is fairly new, some heating contractors and engineers perform blower door testing, others do not. Your state or local energy office might be able to provide sources for the service; also check the telephone book under "Building Inspection."

A blower door test takes about an hour and costs around $75. Often it is done as part of a complete energy audit.

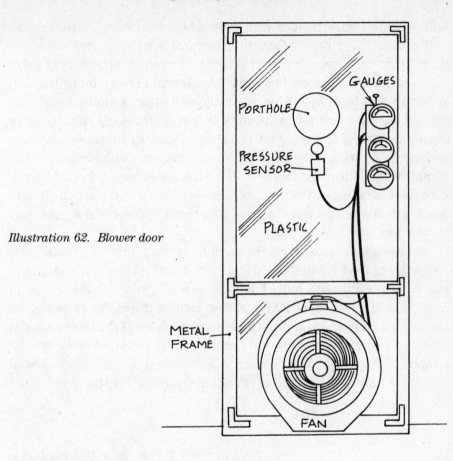

Illustration 62. Blower door

Exhaust Fans

A simple propeller-type, or axial, exhaust fan usually is relatively easy to install in a bathroom or utility room and can eliminate excess moisture in short order. Locating the fan on an outside wall is best; that way long or winding runs of ductwork are not necessary. Another option is to mount the fan in the ceiling with three- or four-inch ductwork leading to the outside (such fans are available that incorporate overhead light fixtures); however, installing these units probably will involve opening the ceiling and/or the floor above to install the ducts.

While bathroom and similar exhaust fans commonly are rated to move air at fifty cubic feet per minute (cfm) to comply with many local building codes, a larger model rated at around 100 cfm usually will provide better results with only a slight increase in power use and noise. The Home Ventilating Institute's recommendations are that bathroom air be ventilated at a rate of eight air changes per hour.

In any case, select a low-noise fan; it will receive more use and generally will be of better mechanical quality than a louder fan. Noise ratings for fans are given in units called sones. One sone roughly equals the sound of a quiet refrigerator; a rating of four sones is twice as loud as a rating of two sones. Some fans are available with timer switches that allow the fan to be left on—for example, to clear moisture from a bathroom after a shower—and then will turn off the fan automatically after a time interval that can be preset and adjusted by the owner.

To install an exhaust fan in an outside wall, first determine how you will connect it to an electrical circuit and where you will mount the switch. Fans with pull chains are not allowed by most codes in areas where moisture is significant—touching the chain while in contact with water could result in an electrical shock if the fan's wiring is faulty (for more on this kind of hazard, called a ground fault, see the chapter "Upgrading Old Wiring"). For fans controlled by a wall switch, mounting the switch near the doorway, on the side by the

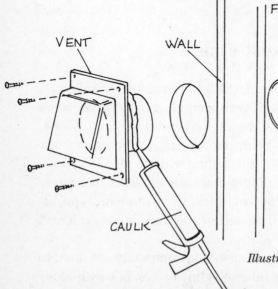
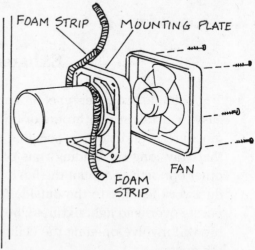

Illustration 63. Installing an exhaust fan

doorknob, is best. Wiring to power the fan usually can run inside the walls from a nearby outlet.

Decide where the fan will go, then cut an opening in the wall large enough for the vent duct but smaller than the fan's mounting plate. Wrap the perimeter of the fan with adhesive-backed foam weather-stripping to seal it and cushion it against the wall, then slide the fan and duct into the opening and secure the fan to the wall with screws. Make the necessary electrical connections following the manufacturer's instructions that come with the fan.

Outside, fill the gap between the sides of the duct and the edges of the hole with fiberglass insulation or urethane foam (this is available in aerosol cans). Attach either a vent hood with a damper or another style of protective cover over the hole and the end of the duct, then test the fan and the cover's moving parts by having a helper operate the fan from indoors. If everything is satisfactory, seal the seams of the cover with caulking compound where the cover touches the wall.

Kitchen Range Hoods

The Home Ventilation Institute's recommendation for kitchen ventilation is fifteen air changes per hour. Virtually the only way to achieve this is by installing a range hood. Range hoods are manufactured in two basic styles: updraft and downdraft. Updraft hoods are installed over a stove and draw cooking exhausts upward. Ordinarily, they are regarded as superior because their effectiveness is enhanced by the natural tendency of hot air to rise. Downdraft hoods draw air downward from the stovetop. They require a more powerful fan than do updraft hoods, and they often are not very effective at capturing steam or cooking exhausts from tall pots or from items placed on front burners.

Both styles of hood can be ducted—that is, connected to a duct that carries exhaust to the outdoors—or recirculating. The latter type draws cooking air through a filter (usually of activated charcoal) to clean it and then releases it again into the room. Ducted hoods are recommended whenever possible; in fact, they are required for gas ranges because recirculating hoods cannot remove combustion products, like carbon monoxide, produced by gas ranges. Consider a re-

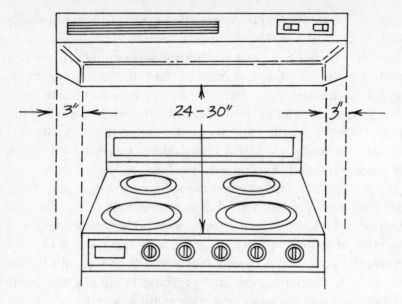

Illustration 63A. Updraft hood

circulating range hood only if installing a ducted hood is impractical or if the house is so tightly constructed that a ducted hood might cause backdrafting. (In the latter case, it is usually worthwhile to have a contractor add draft-assist fans or supplementary fresh-air vents to combustion appliances to augment incoming air supplies.)

The most popular range hoods are conventional updraft models that attach to a wall or beneath a cabinet. Size and height above the cooking area are important: hoods above most stoves should be at least three inches wider on each side than the cooktop and extend far enough forward to cover the front of the stove. A hood for an island- or peninsula-installed stove should extend two to three and a half inches beyond the stove's front and rear edges and three inches beyond each side. The distance between the bottom of the hood and the cooking surface should be between twenty-four and thirty inches for both hoods; increasing the distance up to thirty-six inches is possible but requires a larger hood with a more powerful fan.

Fans for range hoods are either axial (propeller-type) or centrifugal (squirrel-cage-type). Axial fans generally are best for moving large amounts of air at a low speed; they are adequate for hoods having relatively short, straight ducts with few or no bends. Centrifugal fans

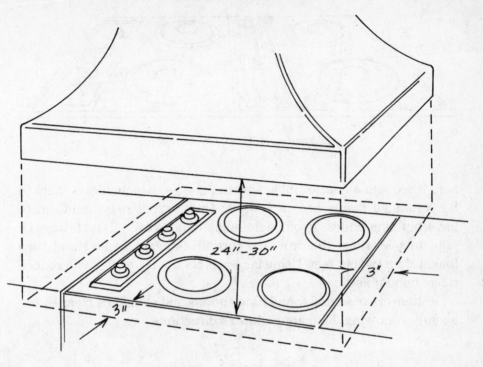

Illustration 63B. Island hood

move air with greater force. As a result, they are well suited to hoods having long and/or circuitous ducts.

Naturally, larger fans move more air than smaller fans. The National Kitchen and Bath Association recommends the following sizes: 300 cfm for a wall- or cabinet-mounted hood over a thirty-six-inch range; 600 cfm for a thirty-six-inch range installed in an island or peninsula counter; 600 cfm for a hood installed over an open grill.

Noise ratings for range hoods, as well as exhaust fans, are expressed in sones. When making a selection, look for fans meeting the recommended size requirements and then compare their sone ratings. For all but the smallest hoods, expect ratings of around six sones. If you are planning a large, custom range hood, investigate roof-mounted fan motors; installing one of these can quiet range hood noise considerably.

Use galvanized or stainless steel ducts. Round or rectangular ducts are acceptable. Standard diameters for round ducts are eight inches, nine inches, and ten inches; standard rectangular ducts mea-

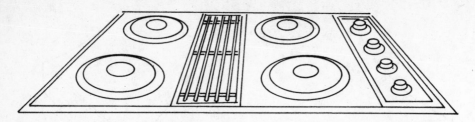

Illustration 63C. Down-draft hood

sure three and a quarter inches by ten inches, which allows them to be contained inside ordinary walls. Duct layout must conform to local building codes. Avoid right-angle bends (use elbow fittings or join duct sections to form two forty-five-degree bends) and runs longer than twenty feet. Using larger ducts than specified can reduce range hood noise.

Filters are essential for all range hoods and should be cleaned frequently, following the manufacturer's directions.

Clothes Dryer Vents

Clothes dryers, whether gas or electric, should be vented to the outdoors to expel moisture. Venting can be through a wall, window, floor, or ceiling. Depending on the dryer's design, the vent can extend from the back of the appliance or from the side. A few models also allow venting from the bottom. For the sake of appearance it is best that the vent's exit be in an inconspicuous location—for example, behind thick shrubbery—because lint as well as moisture is often expelled, and the lint can accumulate in unsightly clumps. However, because the buildup of lint can create a fire hazard under certain conditions, it is important never to terminate a dryer vent under a house, porch, or deck, or in an enclosed airspace such as an air shaft, attic, or chimney. To insure adequate air circulation and to prevent clogging by grass or weeds, the vent opening should be at least a foot above the ground.

Dryer ducts can be rigid or flexible, but should always be metal, not plastic, because of the heat of dryer exhaust. Round ducts, which are standard, should be at least four inches in diameter.

As is true for other ducted ventilation devices, short, straight runs

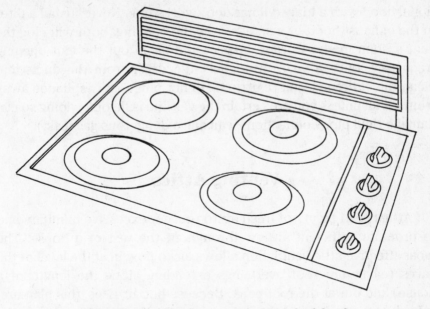

Illustration 63D. Pop-up style down-draft hood

are best. Do not exceed the maximum duct length recommended for the dryer (this information appears in the owner's manual); the blower may lack the power to move air to the outside with enough force to keep moisture and lint suspended. Generally, dryers expel air at the rate of about 180 cubic feet per minute (cfm). As a rule of thumb, a straight run of rigid dryer duct should not exceed thirty feet, assuming that the duct is connected to the dryer by an elbow fitting and that the exit opening is covered by a standard exhaust hood. A similar run of flexible duct should not exceed eighteen feet. With each type, right-angle turns (elbow fittings) must be separated by at least four feet of straight duct. For each added elbow, subtract eight feet from the run's maximum allowable length; no dryer duct should contain more than three right-angle bends.

To assemble duct sections, join them so their crimped ends point toward the outside opening. Use duct tape to seal the seams; avoid screws whose ends can project into the ducts and catch lint. Use a strap-type duct clamp to join the duct to the exhaust opening, and use metal pipe strapping to secure the assembly to walls or framing.

Various exhaust hoods are available. Choose one that is easy to clean and will work despite minor lint buildup. Usually, a conventional hood with a hinged inner door works fine. Never install a filter in the exhaust hood or a screen over the opening; both will clog the vent system in short order. Inspect and clean out the exit opening frequently; once a year, disconnect the dryer from the ducts and clean the entire system. (CAUTION: Do not pull a gas appliance away from a wall unless you are certain the gas line is flexible; doing so can damage rigid pipe connections, causing a dangerous gas leak.)

Venting Attics

Attics need plenty of fresh air to prevent excessive humidity and to promote sufficient airflow throughout the rest of a house. The most effective attic ventilation allows air to flow in at the level of the eaves (eaves are roof overhangs extending along the length of a house) and out at the roof peak. Because hot air rises, this plan creates air movement inside an attic even on a windless day.

Most building codes require the area of attic ventilation openings to equal one square foot for every three hundred square feet of attic space—provided that the attic and living area below are separated by a vapor barrier (for example, insulation batts laid paper face down over an attic floor, or plastic sheeting installed above the ceilings of the rooms below). If there is no vapor barrier, as is often the case, the area of ventilation openings should equal one square foot for every one hundred and fifty square feet of attic space. This means that an attic measuring fifteen hundred square feet should have ventilation openings equaling five square feet if a vapor barrier is installed. Without a vapor barrier, the amount of ventilation area for the same attic must total ten square feet.

Generally, attic ventilation area is divided about equally between eaves-level openings and openings at or near the roof peak. Common and effective eaves-level vents are holes or slots cut in the soffits, which are the undersides of the eaves. These usually are screened to keep out insects and other pests.

Examine soffits to determine whether they contain vents. If they do not, a carpenter, a roofer, or a siding installer can add them. You can do the job yourself in many cases; use a hole saw—an attachment for an electric drill—to bore holes in the soffits where they will be between roof framing pieces. Install screened circular vent plugs,

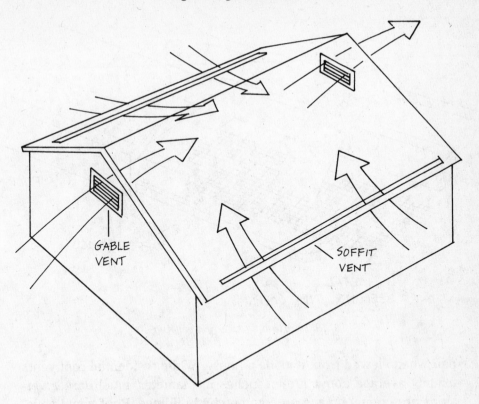

GABLE
VENT

SOFFIT
VENT

Illustration 64. Gable and soffit ventilation

available at hardware stores, by tapping them into the holes from below with a hammer. Soffit-vent kits are also available for creating continuous soffit openings or individual rectangular vents instead of round holes.

In all cases, make certain attic insulation does not cover soffit vents or block the flow of air upward from them along the underside of the roof. Foam and cardboard panels, called baffles, are available to prevent insulation at the sides of the attic from spreading into the soffits. To install, simply place them at the ends of the bays between attic floor joists. If you choose, you can make your own baffles using ordinary cardboard or scraps of lumber.

Many houses have gable vents, louvered openings high up on gable walls. Unfortunately, these generally are too small to be effective, and enlarging them can let in rain and snow. Better solutions are to install roof vents or a continuous ridge vent.

A roof vent is a device that resembles an inverted dish or cooking

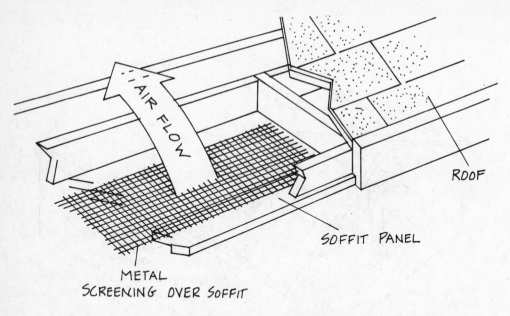

Illustration 64A. Soffit venting

pan when viewed from outside a house. While residential roof vents usually average about fifteen inches in diameter, much larger versions are commonplace on commercial buildings. Roof vents penetrate a roof, allowing air to rise through them. Some models also contain mechanical or electrically powered turbines that revolve, increasing airflow.

Most ridge vents consist of a metal channel that fits over a slot cut along the entire length of a roof peak. The channel must be carefully sealed against the shingles of the roof to prevent leaks; mesh or slots in the underside of the channel permit air rising from inside the attic to escape. Another kind of ridge vent consists of a thick, porous strip instead of a metal channel.

Both roof and ridge vents are effective if properly sized. However, unless you are experienced at roofing carpentry, they should be installed by a qualified roofing contractor to lessen the risk of leaks.

Whole-House Fans

A whole-house fan installed in an attic can draw air from all parts of a house and exhaust it at roof level. If supply air is furnished from

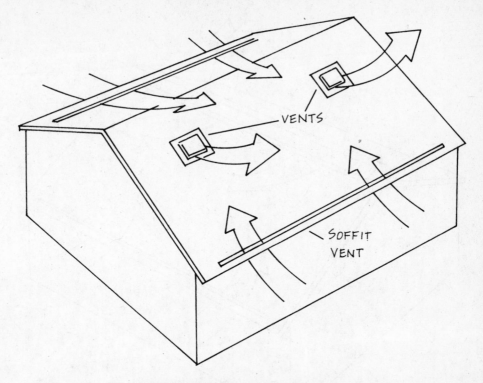

VENTS

SOFFIT
VENT

Illustration 64B. Metal roof vents

near the ground, usually by opening lower windows and basement vents, excellent upward airflow and attendant cooling can be obtained at a cost often averaging 10 percent of the amount required for air-conditioning, albeit without the opportunities for filtration and humidity control.

Standard whole-house fans range in size from twenty-two to thirty-six inches in diameter and cost from around $250 to around $400. Expect to pay between $300 and $350 for a durable model suited to an average-size house. Most units consist of a fan, a mounting frame, an electric motor, and a drive belt that connects the motor to the fan. Some models drive the fan directly, eliminating the need for a drive belt and reducing (so claims the manufacturer) power consumption. Buy a fan with a multispeed motor. Operating the fan at high speed ventilates the house quickly; low-speed operation maintains gentle circulation at a lower noise level after the house cools off.

A whole-house fan should change the air in a house every 2 min-

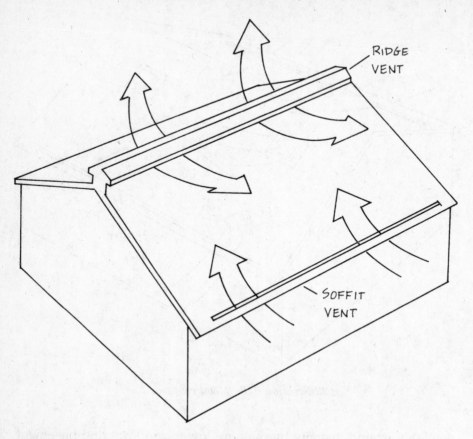

RIDGE
VENT

SOFFIT
VENT

Illustration 64C. Ridge vent

utes. To determine the right size fan to buy, calculate the volume of
all rooms to be cooled (multiply height x width x length of each room
and add the totals), and then divide the total volume by 2. Buy a fan
whose cfm rating is equal to or greater than this amount. (Make sure
the cfm rating is based on 0.1 inch static pressure, not free air. Static
pressure measurements take into account the resistance of any con-
stricting openings through which the air must flow.)

In addition, make sure to provide adequate discharge area for the
fan's exhaust. Attic openings should equal one square foot for each
750 cfm of fan capacity. Louvers and other vents, especially if
equipped with screens for rodent and insect protection, may only be
equivalent to half the area required.

Installing a whole-house fan is a fairly easy job for two people if
at least one of them has moderate home repair experience. However,

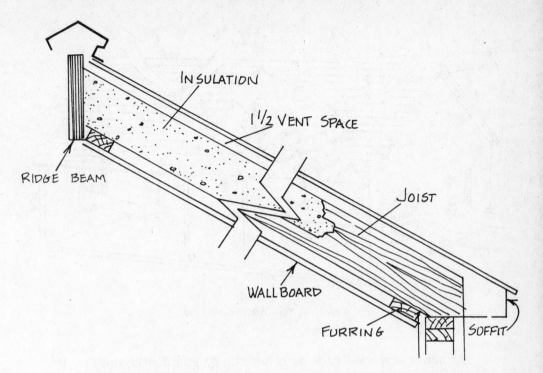

Illustration 64D. Venting a cathedral ceiling

you may need an electrician to install a wall switch conveniently located in one of the rooms and wiring leading to the fan.

Start by determining where to install the fan; usually, the best location is in the ceiling of a central hallway. Examine the attic to make sure there are no obstructions such as electric wires or plumbing pipes in the way, and check to see that there is at least twenty inches of clearance between the tops of the joists on which the fan will rest and the undersides of the roof rafters overhead. From below, mark the outline of the fan on the ceiling (most fans come with a template for this purpose), then saw through the ceiling, following the outline. Be careful not to cut into any of the joists.

Mount the fan by lifting it through the opening and anchoring it with screws to the joists. Rubber mounting pads on top of the joists and rubber washers separating the screw heads from the fan's frame can reduce noise caused by vibration when the fan is running; obtain these from a hardware store or industrial supplier if they are not included with the fan.

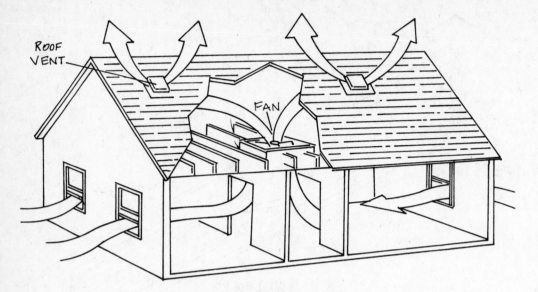

Illustration 65. Whole-house fan

Attach a screened cage or other sturdy guard over the fan. Be sure the guard does not obstruct airflow; while not generally included with a fan kit, a guard is necessary to protect people and stray animals in the attic when the fan is operating.

After making certain the electricity to the circuit governing the fan is shut off, make the electrical connections necessary to operate it. However, do not restore power until installation is complete. From below, install a louvered shutter over the fan opening. Louvers are sold separately and typically cost $50 to $100. Most have spring-loaded vanes that open when the fan is operating and close when the fan is off. When you are finished, restore the electricity and test the fan by operating the wall switch.

During hot weather, operating a whole-house fan is most effective in the evening, after outdoor temperatures have dropped. While it is essential that at least one window or door is open when the fan is running, keeping windows closed in upstairs and unused rooms will increase air velocity through the house.

Heat-Recovery Ventilators

A heat-recovery ventilator, or HRV (the device is also called an air-to-air heat exchanger and sometimes an ERV, for energy-recovery ventilator), is a blower that removes air from indoors and replaces it with outside air. In the process, heat contained in the warmer air—it may be from indoors or outdoors—is transferred to the cooler air. In winter, this exchange preserves indoor heat generated by a furnace or other central heating system; in summer, cooler indoor air results.

All HRVs consist of a cabinet, a core, two fans, and a varying amount of ductwork. Stale air from indoors is drawn by one of the fans into one group of ducts while fresh air from outdoors is drawn into another group of ducts by the second fan. The streams of air pass each other in the core, which is constructed of thin metal or paperlike plastic surfaces that quickly transfer heat. Although the streams of air never actually come into contact with each other, heat carried by the warmer stream migrates through the core and is absorbed by the cooler stream. The heat then remains indoors or is carried outside, depending on the stream's direction.

Heat-recovery ventilators are available in sizes to service individual rooms, larger areas, and entire houses. Installing a whole-house model can save about 70 percent of year-round energy costs that otherwise would be lost because of indoor air's passing unprocessed directly to the outside. However, costs of whole-house units are high, and installation of any HRV requires professional involvement—for selecting, installing, or both. Most whole-house HRVs are installed during new construction and are connected to a separate duct system at a cost comparable to that of an ordinary forced-air central heating loop. While an HRV can be connected to existing central heating and cooling ductwork, this type of installation alone may cost $1,000 to $2,500.

Heat-recovery ventilators are also sometimes used for eliminating radon from basements. For this purpose, smaller units with less ductwork usually are acceptable, and costs often are under $1,000.

Where recovering heat is not necessary, as in an unheated crawl space, a device similar to an HRV but without a heat exchanger can be installed for only a few hundred dollars. Called a balanced exhaust ventilator, or BVS (balanced ventilation system), the appliance sim-

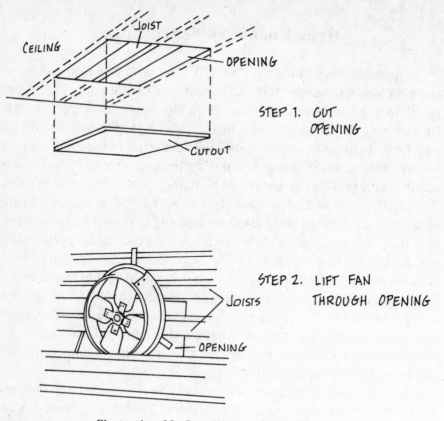

Illustration 66. Installing a whole-house fan

ply removes air from the space in which it is installed and replaces it with an equal amount of outdoor air without altering its temperature.

A BVS, too, can be installed for whole-house use by connecting it to central heating ducts upstream from the furnace. But such an installation requires careful analysis by a heating engineer to make sure the added air does not significantly lower the temperature of the furnace-heated air.

Selecting and sizing an HRV is crucial and should be done by a licensed heating contractor or engineer with HRV experience. However, features of any top-quality device include multispeed or variable-speed blower controls; either a painted, galvanized steel, or sturdy plastic cabinet; easily serviceable parts such as fans and the exchanger core; a defrosting system to prevent internal ice buildup; and UL-listed components. Make sure also that the system will not

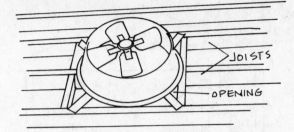

STEP 3. FAN MOUNTED IN PLACE

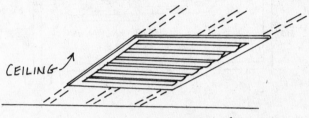

STEP 4. INSTALL LOUVERED SHUTTER

operate for more than fifteen minutes under negative indoor pressure; this prevents backdrafting. The capacity of an HRV—its cfm rating—is more important than heat recovery efficiency. Nevertheless, don't settle for a unit that is less than 65 percent efficient.

Another factor to watch for is how quiet the system is during operation. HRVs can be noisy; they should be installed in a location that is out of earshot (a basement, for example, is a much better home for an HRV than an attic), and special care should be taken to isolate the unit from house framing to minimize vibration noise. Proper duct sizing can have an effect on noise also; avoid short, straight duct runs and ducts that increase low-speed air velocity to more than five hundred feet per minute.

Nearly all HRVs and BVS units produce condensation to some degree and must be installed in a warm location—if necessary, within an insulated compartment—to prevent freeze-ups in winter. HRVs with metal cores produce more condensate than those with plastic cores—plastic cores transfer moisture as well as heat—but whole-house units of both kinds sometimes can lower indoor humidity levels enough to require installation of a humidifier. A final note: intake

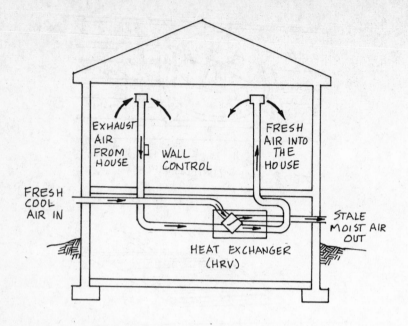

EXHAUST AIR FROM HOUSE

WALL CONTROL

FRESH AIR INTO THE HOUSE

FRESH COOL AIR IN

STALE MOIST AIR OUT

HEAT EXCHANGER (HRV)

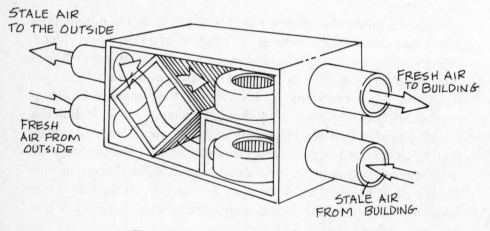

STALE AIR TO THE OUTSIDE

FRESH AIR TO BUILDING

FRESH AIR FROM OUTSIDE

STALE AIR FROM BUILDING

Illustration 67. Heat-recovery ventilator

and exhaust openings for HRVs and BVS units must be separated by at least five feet to prevent exhausted air from being simply recirculated. Also, avoid locating intake openings for any ventilation appliance near pollutant sources such as driveways, carports, animal runs, and areas where garbage cans are kept.

Controlling Noise

What Is It?

SOUNDS ARE PART OF THE NATURAL WORLD. WE ARE SUR-rounded by them; in fact, research has shown that most people feel more comfortable in environments containing steady, low-level, ambient sounds than in surroundings that are completely soundless. But excessive, obtrusive, and unwanted sound is unpleasant and can even produce adverse physical and psychological effects. Acoustics engineers and other sound specialists define this kind of sound as noise, and it can be judged just as much a pollutant as radon, asbestos particles, lead dust, or any of the ubiquitous airborne contaminants discussed elsewhere in this book.

In nonscientific terms, sound is generated by the vibration of molecules; when those molecules that make up solid or liquid materials are set in motion by an impact or impulse, they transmit the force to other molecules nearby. Although the molecules themselves move very little in relation to their original positions, the distance the force passing through them may travel depends on the properties of the other molecules the force encounters. If you imagine a line of billiard balls touching one another, the effect is similar to the way in which a cue ball striking the first ball in line causes only the last ball to move.

Materials made of elastic molecules—these recover their original dimensions quickly after impact—readily transmit vibrations, and therefore sound. By contrast, limp, soft materials absorb vibrations and so do not transmit sound either quickly or fully. As might be expected, hard, solid materials like the wood from which billiard balls are made generally are good sound transmitters. Sound can travel

183

through many varieties of wood, metal, and stone at speeds of up to 20,000 feet per second while diminishing in strength by only one decibel per 3,000 feet. (A decibel, or dB, is a measure of sound theoretically equalling the smallest degree of difference the human ear can perceive; the range of human hearing is about 130 decibels.) Some liquids—for example, water—whose molecules are incompressible also transmit sound well. The speed of sound through water is about 5,000 feet per second.

On the other hand, air is a relatively poor sound transmitter because its molecules compress easily. Not only is the speed of sound through air comparatively low—about 1,100 feet per second—its strength also drops swiftly, by about six decibels for every doubling of the distance between the point at which the sound is measured and its source.

Types of Sound

Two basic types of sound affect building occupants: airborne and structure-borne. Airborne sound is in the form of pressure waves that radiate from their source directly into the air. In a house or other building, airborne sound can exist inside a room or travel along a hallway or through ductwork. When airborne sound strikes a wall or other barrier such as the side of a heating duct, the surface vibrates, transmitting the sound through it and eventually into the air on the other side. Some of the vibrational energy may travel along the surface of the wall or duct and emerge elsewhere in the building, also as airborne sound. Although the wall or duct becomes a secondary radiator of airborne sound and a primary transmitter of structure-borne sound (described below), the classification of the sound overall remains airborne because of its initial source.

Noise caused by airborne sound typically disturbs only those occupants who are relatively near the sound's point of origin. For example, a loud television in one room generally would be heard only by people in adjacent rooms—seldom by those in rooms elsewhere in the house unless they were connected to the TV room by open doors and hallways.

Structure-borne sound is caused by impact. It occurs when building elements such as walls, floors, and ceilings are struck by airborne sound waves or are caused to vibrate by contact with operating appliances or other machinery, or by the movements and activities of occupants. As in the example of airborne sound's traveling through a heating duct, structure-borne vibrations are transmitted through and along surfaces until they eventually radiate as airborne sound; however, their origin—mechanically induced vibration—establishes their classification as structure-borne overall.

The intensity of structure-borne sound usually is greater than that of airborne sound; that is, it usually travels faster and without losing as much strength over long distances. Some structure-borne sounds—for example, those stemming from slamming a door—are of short duration. Others, like vibration from a continually operating fan or heating system, are long-term and can produce adverse effects ranging from simple but ever-present annoyance to the vibrating of dishes in cabinets and glass in windowpanes—and can even lead to structural weakness in extreme cases.

A condition affecting structure-borne sound is the degree to which surfaces amplify vibrations, reinforcing and strengthening them. Like the sounding board of a piano or the body of a guitar, building surfaces that radiate sound do so with an efficiency that varies directly with the ratio of surface area to wavelength or frequency. A water pipe, for instance, radiates comparatively little airborne sound because of its relatively small surface area. This is true even if the source of the sound is large—for example, a vibrating washing machine—in relation to the pipe. However, such a surface will carry high-pitched or high-frequency sound more efficiently than low-frequency sound; therefore, although a water pipe may not transmit the sound of an operating washing machine very far, it may transmit throughout an entire house the sounds of a person turning faucets on or off or striking the pipe with a hammer.

The reverse is true of a large surface like a wall. If a sound source that by itself might generate very little airborne noise is coupled to a large surface so that vibrations—especially those of low frequency—are transferred to it, the intensity or loudness of the sound will be substantially increased. For example, the humming of a refrigerator may be hardly perceptible, but if the appliance is pushed against the wall behind it the motor's vibrations may be amplified by the wall enough to not only create audible noise but also to rattle hanging items and even the floor of a room above.

In the end, most household noise is the result of both airborne and structure-borne sound. The reason is that virtually all items that make noise of their own (airborne sound) also rest on or against floors or walls that acquire and radiate sound vibrations (structure-borne sound). What is more, the ear interprets both types of sound the same way. So, unless a sound can be attributed to a specific source (a piano or a human speaker), there really is no way to tell which is which, and steps to reduce or eliminate the problem must address both.

How Bad Is the Problem?

Few, if any, household noises can damage hearing, but excessive or constant noise can frazzle nerves and interrupt sleep. These conditions in turn can cause psychological distress; in extreme cases they may even lead to physical problems, including high blood pressure and lowered resistance to illness.

COMMON NOISE LEVELS

Sound source	Approximate decibels	Comments
Jet engine	140	
	130	Threshold of pain from sound
"Hard rock" concert	120	
Accelerating motorcycle (50 feet away)	110	
Loud vehicle horn (10 feet away)	100	
Noisy urban street	90	

Sound source	Approximate decibels	Comments
Noisy factory	85	Continuous exposure to sound above this level can cause permanent hearing loss
Noisy school cafeteria	80	
Upper range of speech	72	
Stenographic room	70	
Expressway traffic	60	
Average office	50	
Lower range of speech	42	
Soft radio music in apartment	40	
Average residence (ambient sound)	30	
Average whisper	20	
Rustle of leaves in wind	10	
Human breathing	5	
	0	Threshold of audibility

Getting Rid of Household Noise

It is important to stress again that total absence of sound in a house can be as disturbing to occupants as excessive amounts of noise. That said, there are two general strategies for reducing household noise: halting or containing it at the source, and interrupting its transmission to other areas. Employing elements of both strategies is usually necessary to achieve significant results, but for every household the proportions of each method used and the overall results considered acceptable will vary.

Controlling Noise at the Source

Usually there are only a few noises in a house that can be eliminated or reduced at their sources. Appliances such as clothes washers and dryers, dishwashers, refrigerators, air conditioners, and ventilating fans produce most of these noises; other offenders that can sometimes be silenced are heating system and plumbing parts.

Inspect appliances when you notice any unusual rattle or vibration. Chances are the cause is a loose fastener or other component that can be tightened to eliminate the problem. Also, keep appliances from making direct contact with walls and other surfaces that can magnify what actually may be low noise output. When buying appliances, seek out models rated for quietness. In some cases these contain low-noise motors; in others, extra insulation is used to absorb unwanted sound. Sound ratings for fans are given in units called sones, usually a number from one to ten. The higher the number, the louder the fan; for further information, see the chapter "Combustion Products."

Sound-deadening kits are available for some appliances. Usually, the kits incorporate rubber mounting plates or other parts that separate the appliance from a wall or floor. You can install rubber pads yourself beneath appliances such as washing machines, dryers, and refrigerators. If you cannot obtain sound-deadening kits or pads from an appliance dealer or manufacturer, try an acoustical supply company or an industrial hardware retailer. Both are listed in telephone books.

Broader success is made possible by insulating rooms in which

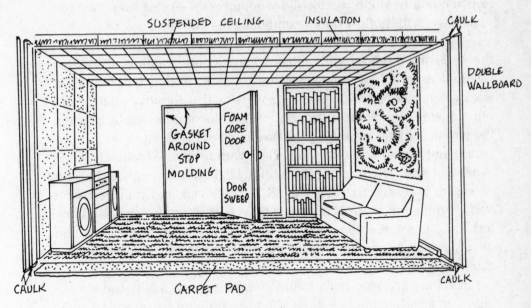

SUSPENDED CEILING   INSULATION   CAULK

DOUBLE WALLBOARD

GASKET AROUND STOP MOLDING

FOAM CORE DOOR

DOOR SWEEP

CAULK

CARPET PAD

CAULK

Illustration 68. Room with moderate sound reduction

noise occurs. Materials such as acoustic tile and thick carpeting, drapes, and wall coverings can absorb airborne sounds such as those made by television and conversation (and, to some extent, those made by appliances or audio equipment isolated from building surfaces) by containing the sounds within the room in which they are produced and by limiting their reflection, which magnifies them. However, these materials will not significantly shield against sounds entering a room from outside, and much of their success depends on so-called leakproof insulation methods; even small openings such as those beneath doors or surrounding ceiling lights and electrical outlets allow sounds to escape an insulated room largely unimpaired.

If your remodeling budget leaves no other choice, sealing openings and adding sound-absorbing materials can be an effective course of action for achieving moderate noise reduction, provided you are thorough and meticulous in your efforts. Begin by installing hollow tubular weatherstripping, also called bulb gasketing, around doorways, and either a doorsweep or a weathertight threshold beneath the door. Thresholds, too, should have a hollow gasket; the shape contains air, which traps sound better than solid material. To be effective, gasketing must be at least a quarter of an inch in diameter.

189

Replace hollow doors with solid wooden ones—or, if these are too expensive, with hollow doors containing insulation.

Also remove any baseboard molding and seal any space between the walls and the floor with silicone caulk (butyl caulk is also effective and usually less expensive; the trade-off is that butyl caulk is stickier and more troublesome to work with). Remove switch and outlet plates from walls (turn off power to the outlets first by tripping the circuit breaker or pulling the fuse controlling them at the service panel) and fill the spaces around them with urethane foam, available in aerosol cans. Do the same with ceiling light fixtures unless they are recessed; recessed fixtures should be removed and replaced with surface-mounted fixtures after you fill the recess with insulation material and patch the opening with wallboard. Finally, search for and fill any remaining openings—for example, holes for plumbing—with caulking compound.

Installing acoustic ceiling tiles—or, better yet, a suspended ceiling—can also help; but again, it will have little effect on sound coming into the room from outside. Panels for suspended ceilings should be at least three-quarters of an inch thick and have an NRC (the initials stand for Noise Reduction Coefficient) rating of at least .65. Fill the space above the panels with unfaced fiberglass insulation at least three and a half inches thick. Thicker insulation can be used if panels are selected that will support the extra weight. (For six-and-a-quarter-inch-thick insulation, which is the maximum recommended, panels one inch thick are required.)

Carpeting must have thick, first-quality padding underneath to be of any use in absorbing structure-borne sound (for example, the sound of footsteps). A pad will also protect the carpet and prolong its life. (Remember, though, that carpet is a potential allergen and VOC hazard. For more on these issues, see the chapters "Volatile Organic Compounds" and "Combustion Products.")

To insulate walls, add a layer of five-eighths-inch-thick wallboard on one or both sides. Attach the wallboard with construction or laminating adhesive instead of screws or nails. (Attaching wallboard with adhesive generally involves fastening the panels temporarily with nails or drywall screws to hold it in place until the adhesive dries. Place strips of wood beneath the fasteners to make removal easier. To install screws, use either an electric drill equipped with a screwdriver bit, or an electric drywall screw gun, which you can buy

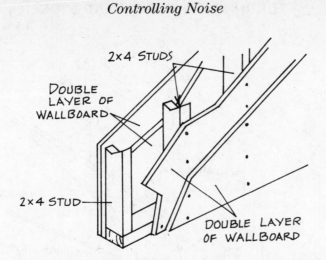

Illustration 68A. Double-wall sound insulation

or rent.) Seal the top and bottom edges of the finished surface with caulking compound; choose a paintable variety unless you plan to install molding to cover it. Another method of attaching multiple wallboard layers is with resilient metal channel or furring strips made for the purpose. This method gives better soundproofing results than using adhesive, but a wall containing strips takes up more space than one without them, and wallboard attached to strips is not sturdy enough to support hanging items like shelves and cabinets. (Mounting such items by using long fasteners that reach all the way to studs in the original wall will defeat the strips' effectiveness.) Installing resilient strips is described in detail on page 193. With both methods, the result should be enough to reduce loud speech heard through the wall to murmers.

Interrupting Noise Transmission

Reducing noise substantially—especially noise caused by impact or emitted as low-frequency vibrations typical of appliances and stereos—requires eliminating structure-borne sounds. This is fairly easy to accomplish in new construction, but remodeling to achieve this goal can be complicated and expensive.

The principle involved is called structural discontinuity, meaning the separation of structural elements so they cannot magnify or transfer sound beyond its source. When structural discontinuity is

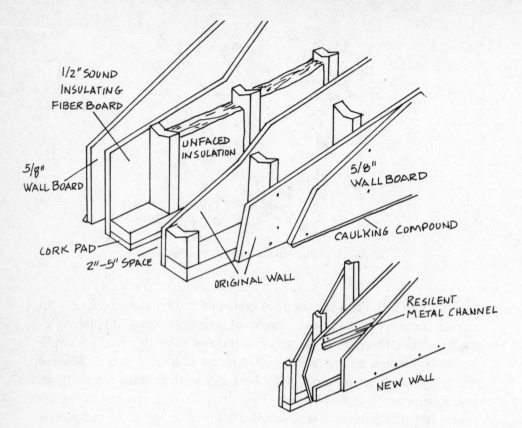

Illustration 68B. Adding a parallel wall

coupled with the sound-absorptive strategies described above, the two techniques together can result in a virtually soundproof environment, with structural discontinuity providing the lion's share.

To isolate noises between two rooms on the same floor, focus your efforts on the wall separating them. The best results usually are obtained by building an additional wall parallel to the dividing wall in whichever room can spare the space. A gap at least an inch wide—although a gap two to five inches wide would be more effective—must exist between the two walls, and it is crucial that no part of them touch. The air between the walls acts as the primary sound absorber; the wider the gap, the better the effect.

The frame of the new wall can be conventionally constructed using two-by-four framing lumber or, better yet, sixteen-gauge two-by-four steel framing pieces; the latter are more resilient than wood and absorb slightly more sound. Install a cork pad made for the purpose (these are available from acoustical supply stores and from some building suppliers) beneath the frame and another above it to separate the frame from the floor and ceiling. In lieu of cork pads, sealing the horizontal edges of the wall frame with caulking compound where the wall meets the floor and ceiling is also effective.

Fill the cavities between the wall studs with unfaced fibrous duct liner or fiberglass insulation. This also will absorb some of the airborne sound passing through the space between the walls. This material may touch the existing wall. (Filling the existing wall with duct liner or insulation will have little additional effect and is seldom worth the work and mess of the demolition involved.) Then attach a layer of half-inch-thick sound-insulating fiberboard (also available at acoustical supply stores and at some building supply stores) to the frame, followed by a layer of five-eighths-inch-thick wallboard. Fasten the wallboard to the fiberboard with adhesive, as was described earlier (page 190), by installing the panels so their seams are perpendicular to those underneath.

Fiberboard and wallboard panels can also be attached with resilient metal channel strips made for the purpose. Resilient strips are somewhat U-shaped in cross-section and create both an airspace and a discontinuous link between the layers. To install strips, fasten them horizontally at 24-inch intervals (measured from the center of one strip to the center of the next) by driving nails or screws through the upper edges of the strips into the wall studs. On a new wall, fasten strips to wall studs before attaching any panels. When modifying an existing wall, attach strips against the existing wallboard. When the strips are in place, fasten panels to them with short wallboard screws (use an electric drill or a drywall screw gun). Avoid driving screws into the underlying wall surface or into studs.

Before finishing the surface of the completed wall with joint compound and paint or other wallcovering, caulk the wall's top, bottom, and sides. It is best if no openings for electrical outlets are installed, but if this is unavoidable, fill areas around the outlet boxes with urethane foam or caulking compound.

If your plans are for new construction—for example, to build walls around a potentially noisy area such as a laundry or workshop, or to build an addition such as a playroom or music room that requires soundproofing—best results are obtained by building double

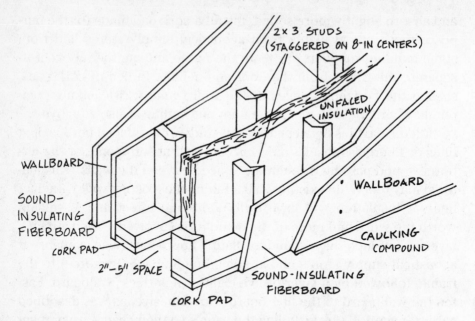

Illustration 68C. Parallel walls: New construction

wall frames for each wall using two-by-three lumber or equivalent sixteen-gauge steel framing. Erect the frames parallel to one another with a gap between them. Fill the cavities between the studs with insulation, then cover the outside of each frame with layers of sound-deadening fiberboard and wallboard. Resilient strips will add some benefit but are not essential.

Constructing sound-resistant walls between rooms may not be enough, as sound can bypass—or flank—them in several ways. For example, sound generated in one room can travel down through the floor and up again through the floor on the wall's other side. Sound also can travel up and over the wall via ceilings and framing.

Because sounds are more effectively contained than repelled, remodeling the floor and ceiling of the noisier room (after separating it behind a sound-resistant wall) is worth considering. Of the two options, remodeling the floor is the most important; unfortunately, it generally also is the most complicated and expensive.

Unless the floor is hardwood or stone, an easy and fairly effective remodel is to remove the finish flooring, cover the subflooring with sound-insulating fiberboard (the same kind used for walls is usually adequate), and then add new finish flooring, preferably carpet or

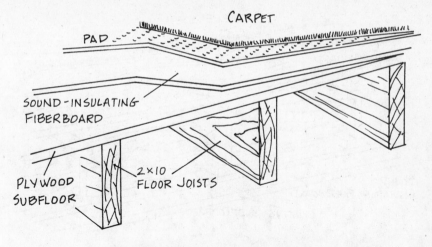

Illustration 68D. Insulated subflooring

other resilient material. (If the existing finish flooring is not easily removed and is in sound condition, sound-insulating fiberboard and new flooring can be installed over it. Of course, doing this will raise the floor level more than if the existing finish flooring is removed.)

A more effective but radical solution that also helps reduce airborne sound is to install a framed floor containing sound-absorbent insulation material above the existing floor. Obviously, a new raised floor will significantly reduce a room's height and might not be a viable solution except in rooms that have high ceilings to begin with. You will also have to step up into the room at doorways. But the method might be easier than removing existing flooring and does have the advantages of providing superior structural discontinuity and allowing installation of hard-surfaced finish flooring.

To install a framed floor, often called a floating floor, start by laying down sound-insulating fiberboard or vibration-isolating rubber strips (available from flooring and building supply companies) directly onto the finish floor or onto the subfloor after the finish floor has been removed. Over this, use construction adhesive to install floor joists of two-by-two or two-by-four lumber, or composite lumber-and-plywood joists designed for flooring. Be sure to use straight joists or laminated ones and bed them firmly to prevent squeaking. Fill the cavities between joists with unfaced fiberglass insulation, then cover the joists with another layer of sound-insulating fiberboard, followed by finish flooring.

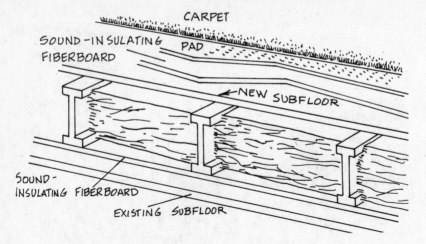

Illustration 68E. A framed, or floating, floor

Ceilings, as was mentioned earlier, can be made to contain sound by installing acoustical tile or a conventional suspended ceiling consisting of a grid and fiberboard panels. Another method is simply to attach resilient channel strips to the existing ceiling (fasten the

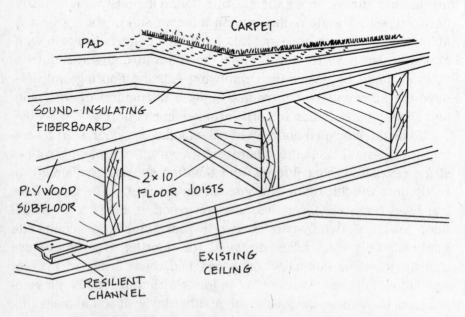

Illustration 68F. Suspending a ceiling with resilient strips

196

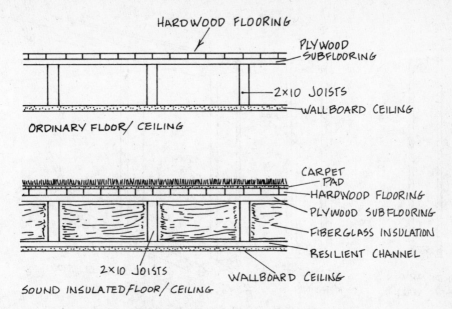

HARDWOOD FLOORING

PLYWOOD SUBFLOORING

2x10 JOISTS

WALLBOARD CEILING

ORDINARY FLOOR/CEILING

CARPET
PAD

HARDWOOD FLOORING

PLYWOOD SUBFLOORING

FIBERGLASS INSULATION

RESILIENT CHANNEL

2x10 JOISTS

WALLBOARD CEILING

SOUND INSULATED FLOOR/CEILING

Illustration 68G. Floor/ceiling insulation methods

strips through the ceiling material into the joists) and install a second layer of wallboard against them to create a new, discontinuous ceiling.

Insulating an upstairs room from the downstairs room beneath it (or vice versa) requires remodeling the floor/ceiling assembly separating them by combining the methods described above; for example, by installing a suspended ceiling in the downstairs room and a floating floor in the room above. Again, because sound is more effectively contained than repelled, the secret of success lies in focusing attention on the noisier area.

Other Methods

The drawings on these pages show a variety of wall, floor, and ceiling assemblies, each rated according to its average resistance to sound by conventional acoustics engineering standards. STC ratings—the initials stand for Sound Transmission Class—evaluate materials according to how well they resist airborne sound; an STC rating number equals the number of decibels a sound is reduced as it passes through a material or assembly. IIC ratings—the initials stand

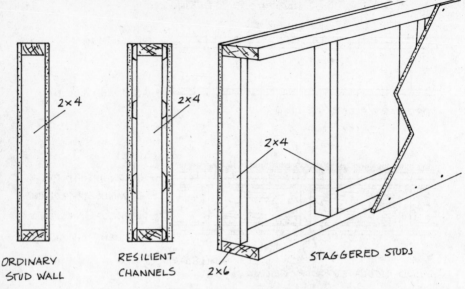

2×4

2×4

2×4

ORDINARY
STUD WALL

RESILIENT
CHANNELS

2×6

STAGGERED STUDS

Illustration 68H–J. Other ways to impede sound

for Impact Insulation Class—evaluate materials and assemblies according to their resistance to impact sound. For both standards, higher numbers indicate better resistance than lower numbers.

Ordinary partition walls made of wooden studs with a single layer of wallboard fastened directly to their surfaces have an STC rating of around 35; dramatic improvement is noticed when ratings are between approximately 40 and 65. When selecting a sound-resistant wall design, remember that if the wall is erected parallel to an existing wall, the total STC rating will equal or exceed the combined ratings of both walls due to the insulating qualities of the airspace between them. Closets and well-stocked floor-to-ceiling bookcases can also act as buffers against noise. The ratings shown with the drawings imply careful construction, caulking around the perimeter, and no openings for electrical outlets, heat registers, doors, or windows.

Coping with Outdoor Noise

Noise enters houses from outdoors through the same openings that admit air; therefore, the same steps that "tighten" a house to

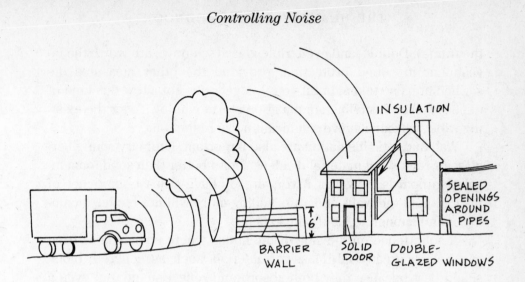

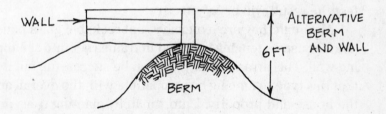

Illustration 69. Remedies for outdoor noise

make it more energy-efficient will also help eliminate unwanted sounds.

Chief among noise-reducing steps is sealing exterior openings of all sizes, from cracks to gaps, with caulking and high-quality bulb-type weatherstripping, described earlier in this chapter. Pay particular attention to the seams around windows and doors; openings for plumbing, wiring, and exhaust vents; and places where the house framing rests on the foundation.

Solid doors insulate against sound better than hollow doors, and single-panel solid doors insulate better than frame-and-panel doors (the latter allow air to pass through the seams where the panels join the frame). Double- and even triple-glazed windows are worth the investment in sound protection, provided the other measures described are also taken; but if your budget does not allow this kind of expense, even installing combination storm windows over the existing windows can achieve significant noise reduction.

Wall and attic insulation are also important. Ordinary spun fiberglass or rock wool material deadens sound better than solid foam insulation boards; however, if you already have at least six inches of insulation in the walls and attic, adding more will not reduce incoming sound enough to justify the expense.

A solid fence on the side of the house where noise comes from can help a great deal. Massive materials such as concrete block shield best because they both absorb and reflect sound; but even a wooden fence, provided it is solid (not a picket fence), can repel most street noise, with the exception of large trucks. Fences should be at least six feet high to protect against traffic noise. An alternative to this is to create a mound of earth, called a berm, and add a fence at the top. As long as the height of the combined construction totals six feet the result will be as effective.

Although they are expensive, unbreakable glass panels are available for use as fencing installed in frames set into the top of a retaining wall. This arrangement, too, can be successful, but it is important that this type of fencing be compatible with the overall appearance of the house and property. Unfortunately, planting trees and hedges to protect a house from noise does little good except from a psychological viewpoint. However, studies have shown that sounds are less annoying when their sources are invisible.

Plumbing and Heating System Noises

Household plumbing and central heating and cooling systems often create annoying noise. Most are the result of incorrect installation or the need for repair or adjustment. As mentioned earlier, all openings where pipes penetrate walls, floors, and ceilings require sealing with caulking compound or polyurethane foam to prevent noise leaks. Walls containing pipes should be completely finished to floor level and sealed at the top and bottom.

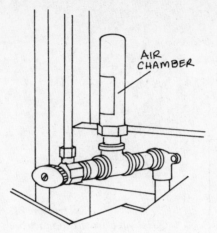

AIR
CHAMBER

Illustration 70. Air chamber

Plumbing Noises

To diagnose noisy plumbing, it is important to determine first whether the unwanted sounds occur on the system's inlet side—in other words, when water is turned on—or on the drain side. Noises on the inlet side have varied causes: excessive water pressure, worn valve and faucet parts, improperly connected pumps or other appliances, incorrectly placed pipe fasteners, and plumbing runs containing too many tight bends or other restrictions. Noises on the drain side usually stem from poor location or, as with some inlet side noise, a layout containing tight bends.

Hissing Hissing noise that occurs when a faucet is opened slightly generally signals excessive water pressure. Consult your local water company if you suspect this problem; it will be able to tell you the water pressure in your area and can install a pressure-reducing valve on the incoming water supply pipe if necessary.

Thudding Thudding noise, often accompanied by shuddering pipes, when a faucet or appliance valve is turned off is a condition called water hammer. The noise and vibration are caused by the reverberating wave of pressure in the water, which suddenly has no place to go. Sometimes opening a valve that discharges water quickly into a section of piping containing a restriction, elbow, or tee fitting can produce the same condition.

Water hammer can usually be cured by installing fittings called air chambers or shock absorbers in the plumbing to which the problem

201

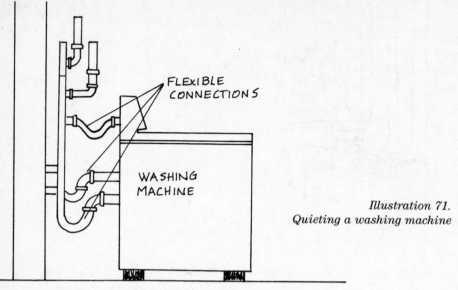

FLEXIBLE CONNECTIONS

WASHING MACHINE

Illustration 71.
Quieting a washing machine

RUBBER BLOCKS

valves or faucets are connected. These devices allow the shock wave created by the halted flow of water to dissipate in the air they contain, which (unlike water) is compressible.

Older plumbing systems may have short vertical sections of capped pipe behind walls on faucet runs for the same purpose; these can eventually fill with water, reducing or destroying their effectiveness. The cure is to drain the water system completely by shutting off the main water supply valve and opening all faucets. Then open the main supply valve and close the faucets one at a time, starting with the faucet nearest the valve and ending with the one farthest away.

Chattering or Screeching Intense chattering or screeching that occurs when a valve or faucet is turned on, and that usually disappears when the fitting is opened fully, signals loose or defective internal parts. The solution is to replace the valve or faucet with a new one.

Pumps and appliances such as washing machines and dishwashers can transfer motor noise to pipes if they are improperly connected. Link such items to plumbing with plastic or rubber hoses—never rigid pipe—to isolate them.

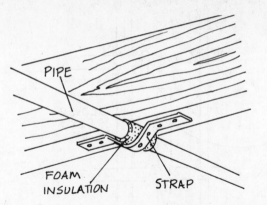

*Illustration 72.
Securing pipe
perpendicular to joist*

Other Inlet Side Noises Creaking, squeaking, scratching, snapping, and tapping usually are caused by the expansion or contraction of pipes, generally copper ones supplying hot water. The sounds occur as the pipes slide against loose fasteners or strike nearby house framing. You can often pinpoint the location of the problem if the pipes are exposed; just follow the sound when the pipes are making noise. Most likely you will discover a loose pipe hanger or an area where pipes lie so close to floor joists or other framing pieces that they clatter against them. Attaching foam pipe insulation around the pipes at the point of contact should remedy the problem. Be sure straps and hangers are secure and provide adequate support. Where possible, pipe fasteners should be attached to massive structural elements such as foundation walls instead of to fram-

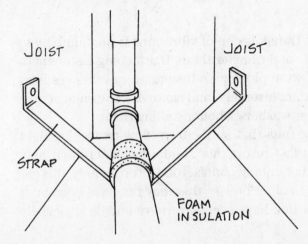

*Illustration 72A.
Securing pipe
parallel to joist*

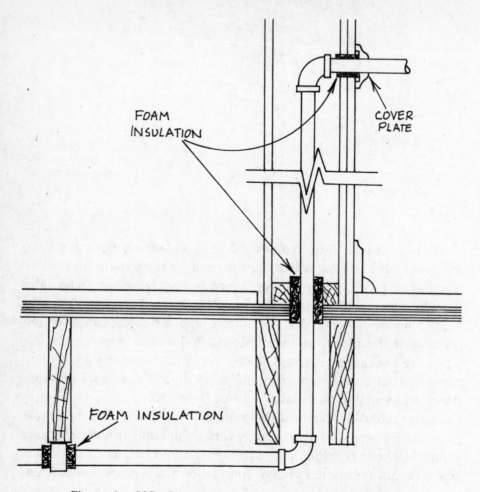

FOAM
INSULATION

COVER
PLATE

FOAM INSULATION

Illustration 72B. Silencing a pipe run through floor and walls

ing; doing so lessens the transmission of vibrations from plumbing to surfaces that can amplify and transfer them. If attaching fasteners to framing is unavoidable, wrap pipes with insulation or other resilient material where they contact fasteners, and sandwich the ends of new fasteners between rubber washers when installing them.

Correcting plumbing runs that suffer from flow-restricting tight or numerous bends is a last resort that should be undertaken only after consulting a skilled plumbing contractor. Unfortunately, this situation is fairly common in older houses that may not have been built with indoor plumbing or that have seen several remodels, especially by amateurs.

Drainpipe Noise   On the drain side of plumbing, the chief goals are to eliminate surfaces that can be struck by falling or rushing water and to insulate pipes to contain unavoidable sounds.

In new construction, bathtubs, shower stalls, toilets, and wall-mounted sinks and basins should be set on or against resilient under-layments to reduce the transmission of sound through them. Water-saving toilets and faucets are less noisy than conventional models; install them instead of older types even if codes in your area still permit using older fixtures.

Drainpipes that do not run vertically to the basement or that branch into horizontal pipe runs supported at floor joists or other framing present particularly troublesome noise problems. Such pipes are large enough to radiate considerable vibration; they also carry significant amounts of water, which makes the situation worse. In new construction, specify cast-iron soil pipes (the large pipes that drain toilets) if you can afford them. Their massiveness contains much of the noise made by water passing through them. Also, avoid routing drainpipes in walls shared with bedrooms and rooms where people gather. Walls containing drainpipes should be soundproofed as was described earlier, using double panels of sound-insulating fiberboard and wallboard. Pipes themselves can be wrapped with special fiberglass insulation made for the purpose; such pipes have an impervious vinyl skin (sometimes containing lead). Results are not always satisfactory.

Noisy Well Water

If your water comes from a well, check the pressure yourself by inspecting the gauge on the water storage tank. The pressure should not exceed sixty pounds per square inch (psi). Most storage tanks for well systems are regulated by a pressure switch and an air-volume control. The switch permits the pump to operate. As water enters the tank it pushes air above it into the tank's upper part; when the tank fills to a level at which the pressure of the compressed air is between 50 and 60 psi, the switch responds by shutting off the pump. As water then is drained from the tank by use, the pressure eventually drops to between 30 and 40 psi, which again triggers the switch, reactivating the pump.

Pressure tanks also have air-volume controls that regulate the amount of air tanks contain. Incoming water can admit small amounts of air that eventually accumulate, causing tank pressure to rise prematurely. The volume control assures that extra air is released.

If opening faucets produces a sharp blast of air, the pressure tank may be air-bound thanks to a faulty air-volume control; that is, the tank may contain too much air and not enough water even though the pressure inside the tank is normal. To remedy the situation, shut off power to the pump, open the drain valve at the bottom of the tank, and let the water inside run out until the pressure gauge reads zero (attach a garden hose to the valve to direct the water to a floor drain if necessary). Replace the air-volume control with a new one. Then close the drain valve and turn on the pump to refill the tank.

Some pressure tanks, usually small ones, have an inner balloon or diaphragm that separates the air in the tank from the water. Consistent low water pressure, which causes the pump to turn on and off frequently, can mean a leak in the diaphragm. However, symptoms of an air-bound tank usually indicate a leak in the piping between the well and the tank. Call a well contractor if you cannot find the leak and repair it.

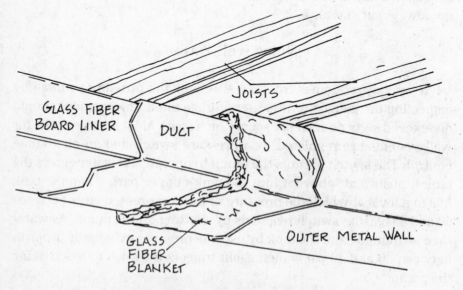

Illustration 73. Double-walled duct

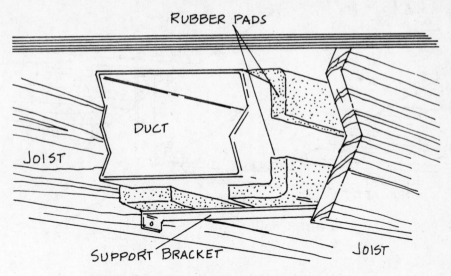

RUBBER PADS

DUCT

JOIST

SUPPORT BRACKET

JOIST

Illustration 73A. Mounting duct parallel to joists

Heating and Ventilation System Noise

Noisy heating and cooling systems, and exhaust fans, usually are the result of incorrectly installed units that transmit motor noise via house framing and ducts. Sometimes air moving at high velocity through ducts is also to blame.

Equipment Noise Furnaces, central air conditioners, blowers, and similar units should be professionally installed and repaired. Mountings for such units should include rubber or other resilient blocks that dampen vibration. Exhaust and whole-house fans can be installed by do-it-yourselfers, but similar mounting methods should be employed.

When selecting heating and ventilation equipment, it pays to invest in quiet-rated models and designs. Generally speaking, oversized equipment (which has greater capacity than required) is quieter to operate than smaller equipment, which must labor continuously at maximum output to produce the same results. However, selecting equipment that is too large can also create excess noise due to extreme air velocity; always consult a heating engineer or knowledgeable contractor before purchasing. Large-diameter, slow-moving belt-driven blowers are less noisy than smaller, high-speed blowers

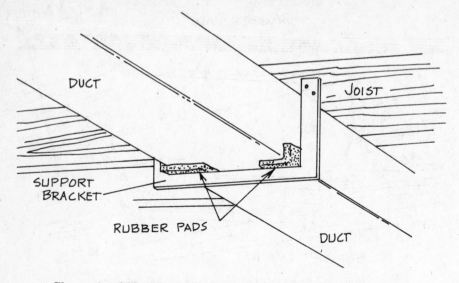

Illustration 73B. Mounting duct running perpendicular to joists

coupled directly to motors. Air passing from blowers into ductwork should first enter a plenum, or calming chamber, which is a large enclosure that allows turbulence in the air to subside. To bar vibration, the chamber should be connected with flexible rubber boots both to the blower and to the duct entrance.

Centrifugal, or squirrel-cage, fans generally are quieter than vane-axial, or propeller-style, fans. All massive heating and air-conditioning equipment should be mounted away from living areas, preferably in the basement or outdoors. Avoid installations in utility rooms, closets, and attics, and on rooftops.

Duct Noise Ductwork requires careful selection, layout, and installation. Size should be determined by a heating engineer or contractor, and all ducts should be lined with acoustic insulation (fibrous duct liner). Double-wall, acoustically insulated ducts are best. Ducts must be mounted securely, away from contact with structural framing, using resilient hangers and duct supports that will not transmit vibration. Where ducts pass through floors and ceilings, they should be protected by resilient sleeves or collars unless the gaps are wide enough that contact with the sides of the openings is not possible. Exposed ductwork of all kinds can be further soundproofed by enclosing—or boxing—it with sound-insulating fiberboard and wallboard fastened to a framework of two-by-twos or other lightweight furring strips. Boxing is described and illustrated on pages 16 and 17.

Noise can also occur inside ducts at joints between sections, at sharp bends, and where branches of ductwork join. Turbulence is the most frequent cause, even if a plenum chamber is installed. To remedy the condition, make sure no sharp edges project inside ductwork at joints, and that all seams between sections are sealed with duct tape. Baffles and damper plates can be installed at ell and tee connections to dissipate turbulence where otherwise an airstream would strike the sides of ducts with full force.

Grille Noise   A high-pitched whistling sound at the entrance of a duct into a room can be caused by the grille, or register, covering the duct's opening. As a test, remove the grille. If the sound disappears, the grille is the cause. Replace the grille with a sturdier model whose louvers do not vibrate, or else install a larger grille that restricts less air. Doubling the area of a grille lowers the velocity of air passing through it by 50 percent; this can reduce resulting noise by as much as twenty-five decibels.

Location of grilles is also important. Do not install grilles in corners where nearby walls can capture and project sound into the room. If an existing grille is in such a location, consider moving it if it is noisy. The ideal location for duct openings is at least six feet from corners.

PART II

Minimizing Safety Hazards

Protecting Your Home
from Fire

FIRE IS A HOME'S DEADLIEST AND MOST DEVASTATING HAZard. According to estimates by the National Fire Protection Association, over two thousand homes in the United States are struck by fire every day, an average of one fire every fifty seconds. Four thousand to five thousand Americans die in fires each year, according to the Consumer Product Safety Commission (CPSC), and tens of thousands are injured. Annual property damage is in the billions of dollars.

Protecting a home and its occupants from fire requires three things: eliminating fire hazards, installing an alarm system (smoke detectors), and practicing how to respond in case a fire does start. Among the leading causes of home fires are kitchen accidents, faulty household wiring, faulty heating equipment and appliances, and improperly stored flammable materials. All are preventable, often easily and at little cost, but the key is to act promptly; the sooner you take steps to safeguard your home against fire, the less you risk your own life and the lives of family members and other occupants.

To locate fire hazards, tour your home with an eagle eye for them. Be sure to inspect the attic, basement, garage, and other areas that might otherwise be overlooked. Remedy as many hazards as you can on the spot, and fix others as promptly as possible or have them fixed by a professional. In a large home, make several passes, each with a different focus, instead of trying to locate and eliminate all hazards room by room.

Inspecting for faulty electrical outlets is one place to start. (Note that only fire hazards associated with electrical systems are covered

213

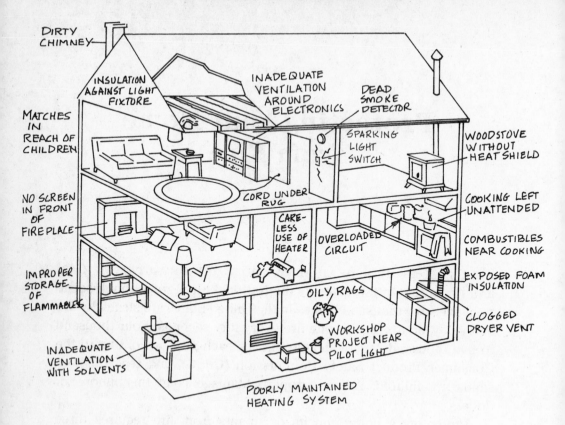

Illustration 74. Household fire hazards

in this chapter. Inspecting for shock hazards is covered in the chapter "Avoiding Falls and Other Injuries"; for complete details on evaluating and repairing household electrical wiring and outlets, see the chapter "Upgrading Old Wiring.")

The most obvious electrical outlet hazards to look for are overloaded receptacles, which means there are too many items plugged into wall outlets. Of course, if an overload is significant it will result in a blown fuse or tripped circuit breaker; regard such an event as a warning signal and take the steps described below to remedy the problem. But if an overload is only marginal, the fuse or breaker monitoring the circuit may not be triggered; this allows wiring gradually to deteriorate due to excessive heat over long periods of time—and the result can be a fire.

Overloaded circuits generally occur when plug-in extenders are used to accommodate more appliances than the outlet is designed to accept. The best solution is simply to reduce the number of appliance connections to a receptacle by moving some items to other receptacles or by having additional receptacles installed. If this is not feasible, invest in strip-type extenders that have a self-contained fuse and a single cord that plugs into the wall receptacle. Strip extenders rest on the floor, or they can be mounted on the wall above a counter. The fuse of the extender is more sensitive to overload than the fuse or breaker governing the circuit; therefore, if too many appliances are plugged in, the extender's fuse blows before the circuit wiring overheats. An additional advantage is that receptacles are not overcrowded by numerous plugs that can be kicked loose or slip by their own weight.

Inspect receptacles also for cracks, scorched or blackened areas, and the inability to hold plugs tightly. If you spot any of these flaws, or if the faceplate feels warm to the touch, have the receptacle replaced. As for appliance cords and plugs, these, too, require replacing if they are cracked, frayed, or burned. Feel the plugs of appliances while they are operating; slight warmth can be expected, but if a plug is hot to the touch, turn the appliance off and unplug it. Have the appliance inspected professionally before using it again; one possibility is that the cord is undersized and must be replaced.

Remove appliance and extension cords from beneath rugs or furniture where they can be abraded and start a fire. Never attach extension cords to walls, baseboards, or elsewhere for use as permanent wiring; if you live in an old or remodeled house having such installations, hire a licensed electrician to replace them with new circuit wiring that meets the standards of the electrical code in your area. (Temporarily fastening appliance and extension cords to walls or baseboards is a good idea if your aim is to prevent children or pets from disturbing them, but use strong tape instead of metal fasteners that can penetrate cords; such mishaps can cause dangerous short circuits.) When using an extension cord with a heavy-duty appliance or with power tools, be sure the cord is rated to accommodate increased electrical loads. Ordinary household extension cords are not sufficient.

Older receptacles may contain sockets having only two slots each instead of three. The differences between two- and three-slot sockets and the hazards associated with misunderstanding and misusing them are fully described in the chapter "Upgrading Old Wiring." Suffice it to say here that two-slot sockets are not safe for three-prong appliance plugs, even if you equip the receptacles with plug-in adapters supposedly designed for the purpose. The "extra" prong on three-prong appliance cords is a safety feature designed to connect with a grounding terminal inside receptacles that is itself linked to a grounding system protecting the entire circuit. In most cases, the presence of receptacles with two-slot sockets indicates the absence of a circuit-grounding system in the first place. So, although properly installing a plug-in adapter will effectively ground the appliance cord's third prong to the receptacle (actually, to the metal box, located inside the wall, containing the receptacle), neither the receptacle nor the box is grounded; therefore, the adapter provides no real safety from shock or overheating.

Inspect all wall switches to make sure they function. Any that feel warm to the touch, spark, or do not work must be replaced. Also inspect ceiling light fixtures for scorched or burned areas and brittle or flaking wires; all are evidence of overheating. Replace bulbs having higher wattages than the fixtures' rating labels allow with bulbs having wattages that are the same or lower than allowed. If you do not know the correct wattage for a light fixture, use a bulb no larger than sixty watts.

In attics, check that recessed light fixtures in the ceiling below are not covered by insulation unless they are rated "IC" (insulated ceiling). Ordinary recessed fixtures must be separated from insulation by a gap of at least three inches on all sides. Boxes built around recessed fixtures to support insulation above them are not recommended either, even if the insulation remains three inches away. Check light fixtures in closets to make sure all bulbs are shielded from contact with items on shelves. If you are planning to build a closet, check local building and electrical code regulations beforehand to learn safe locations for closet lighting fixtures.

Be sure heating equipment appliances such as electric baseboard heaters, space heaters, fireplaces, and woodstoves are adequately shielded from contact with nearby surfaces and are away from furnishings. Maintain heating equipment diligently. Electric baseboard heaters should be vacuumed and inspected for faulty wiring often. Electric space heaters, too, should be examined often for frayed cords, broken or loose wiring, and overheating wall plugs.

Fireplaces and woodstoves require similar vigilance. Inspecting fireplaces and using them safely is covered in the chapter "Combustion Products." If you uncover a fireplace during remodeling or are thinking of activating a fireplace that has been unused for years, have it professionally examined. When using a fireplace, make sure a fire screen is in place to prevent sparks from flying out. Such a screen should fit closely over the entire fireplace opening. Even so, never store newspapers, kindling, or firewood near a hearth—and *never* leave a fire unattended.

Woodstoves must be installed according to local building code regulations. If you live in a home with a woodstove that was previously installed, have it inspected by a building inspector or fire marshal. If you buy a new stove, have it installed by a certified installer. Beware when buying a used stove; very likely it does not meet current EPA emissions standards (for more on these, see the chapter "Combustion Products.")

All woodstoves must be placed no closer to combustible surfaces—walls, ceilings, floors, and furniture—than established minimum clearances permit. These clearances vary depending on stove design and types of shielding used with them to protect combustible surfaces from heat. Standards normally applied by building codes and stove manufacturers are those developed by Underwriters Laboratories (UL) and the National Fire Protection Association (NFPA).

Combustible surfaces include wood-framed walls, ceilings, and floors covered with noncombustible materials such as gypsum wallboard, plaster, and brick. Heat from a stove can pass through such noncombustible materials and ignite the combustible lumber on the other side. In all cases, no woodstove lacking heat shield design features can be closer than thirty-six inches from an unprotected combustible wall or ceiling, and all woodstoves must rest on noncombustible hearth materials to protect a combustible floor.

Heat-reflecting panels of sheet metal or cementitious backer board (wallboardlike material used as a base for ceramic tile installations) can be attached to walls for protection, thereby reducing required clearance, but such shielding must be attached in a way that maintains an airspace of at least an inch between it and the wall. One way to do this is to use porcelain electrical insulators as separators. Hearths generally must be more than simply a layer of brick bedded in sand or mortar; unless the stove has a bottom heat shield, a layer

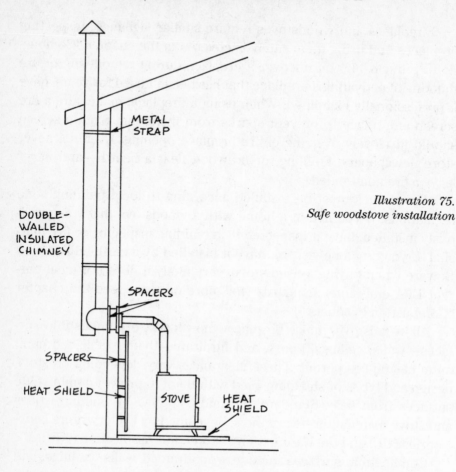

Illustration 75.
Safe woodstove installation

of insulation (two thicknesses of backer board usually is adequate) must separate these materials from any combustible materials underneath. Wall shield kits and other protective paraphernalia for stoves are available at fireplace and woodstove stores.

Just as important, have the chimney or stovepipe inspected and cleaned at least once a year—or more often if you build fires regularly. Cleaning a chimney or stovepipe in the fall, midway through the winter, and again in the early spring is a good plan if you live in a northern region and keep a fire going constantly during the cold months. (For more on chimney safety, see the chapter "Combustion Products.")

Kerosene-burning space heaters are extremely dangerous and whenever possible should not be used. Local fire codes in many areas prohibit them. If you do use them, carefully follow the operating and maintenance instructions supplied by the manufacturer; make sure their tip-over safety switches work; keep heaters at least three feet from combustible surfaces, including drapes, tablecloths, and both upholstered and solid wood furniture; never leave an operating heater unattended; and always refuel a heater outdoors.

Fuel cans for heaters should be colored differently from any you might use for other flammables and should be used only for kerosene. Burn only number-one grade, "water-clear" kerosene in heaters; never use lesser grades that appear yellow. Naturally, never attempt to use gasoline or other flammable liquid as fuel—an explosion is virtually certain.

In the kitchen, inspect cooking surfaces and nearby walls and cabinets for grease buildup (clean such areas by scrubbing with ammonia, followed by detergent and water), and check to see that curtains, napkins, potholders, and aprons are kept far away from burners and heat. Do not store newspapers or paper bags near the stove. Check that major appliances such as the refrigerator and dishwasher are in working order and keep them in top condition by cleaning them regularly and having them periodically maintained by a qualified service technician. Keep small appliances clean, dry, and functioning properly; always unplug them when they're not in use.

Avoid performing refinishing projects or setting up a home workshop in an area containing a pilot light, even if it is across a room or otherwise relatively far away. Fumes and sawdust from project materials can travel surprising distances and ignite or explode in the presence of an open flame. No matter where you locate a shop, do not keep flammable liquids indoors. Instead, store them in ventilated, locked safety cabinets—preferably made of metal—in a garage or shed separate from the building. Store flammables in an attached garage only if partition walls and doors separating it from the house are fire-resistant—a requirement of most building codes. Promptly dispose of rags and paper containing flammable substances; these can spontaneously ignite in closed containers or even when subjected to sunlight filtering through a window.

Selecting and Installing Smoke Detectors

Smoke detectors are the first line of defense against fire. According to the National Fire Protection Association, having working smoke detectors reduces by half your chances of dying in a home fire—because at least half of all fatal home fires in the United States occur at night, when occupants are asleep. A smoke detector can awaken you and the other members of your family in time to prevent suffocation (the cause of at least two-thirds of all fire-related fatalities) and get you out of the house alive.

There are two main types of smoke detectors. Photoelectric detectors generate a beam of infrared light; smoke interrupting the beam causes the alarm to sound. Photoelectric detectors are sensitive to smoldering fires in upholstery, in bedding, and behind walls. As a result, they are well suited for installation near living and sleeping areas. Unfortunately, photoelectric detectors also respond to steam, dust, and other airborne particles; therefore, they are not a good choice for bathrooms, laundry rooms, or home workshops.

To protect those areas, install ionization detectors. These emit a minute amount of radiation that electrically charges, or ionizes, the air inside the device; smoke entering the detector reduces the electric current, activating the alarm. Ionization detectors are most sensitive to fast-developing fires from highly flammable materials such as paper, wood, and refinishing products, and to fires that give off little smoke. A third type of smoke detector combines both technologies.

For kitchens, photoelectric detectors are better than ionization models because the latter tend to sound accidentally in the presence of cooking smoke. Best of all, however, is a heat detector; these respond to changes in temperature, not to the presence of smoke or fumes. (Avoid using heat detectors elsewhere; because high heat is required to activate them, a fire could be well under way before the alarm sounds.)

Some smoke detectors offer other features as well. At least one brand can be tested merely by shining a flashlight at a particular spot, thereby eliminating the need to climb a footstool or ladder to press a button. Other models can be deactivated for up to fifteen minutes to allow cooking or lighting a fireplace. Still others contain emergency lighting or can be connected to a whole-house security system. Re-

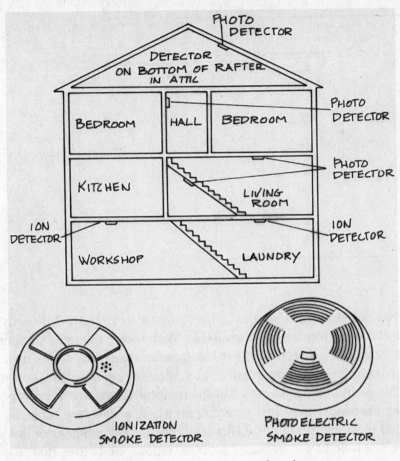

Illustration 76. Smoke detector locations

cently, smoke detectors for the hearing-impaired have become available; these are equipped with a powerful strobe light that can awaken sleeping occupants with their bright flashes.

The least expensive smoke detectors, and the easiest to install, are battery-powered. Of course, they are useless if the battery is allowed to run down or is removed (for example, to power another device or to silence false alarms). Manufacturers recommend changing smoke detector batteries once a year—for instance, on the day clocks are turned back each fall from daylight saving to standard time. (Lately, one manufacturer has introduced a smoke detector with a lithium battery advertised to last six to ten years. The unit "chirps" periodically for 30 days to warn that the battery is failing.)

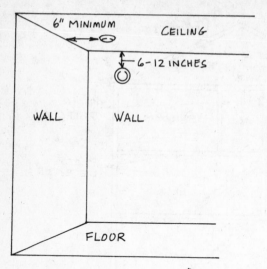

6" MINIMUM CEILING

6-12 INCHES

WALL   WALL

Illustration 76A.
Smoke detector placement

FLOOR

Smoke detectors that can be plugged in or wired directly to household current are also available. With these, the need to change batteries is eliminated, and for that reason many fire codes now require them for new construction, including remodeling. Besides eliminating maintenance hassles and the temptation to use the battery for other purposes, they are convenient when a detector must be installed in a location that is difficult to reach. Also, whether or not you have a whole-house security system, smoke detectors that are directly connected to household wiring can be interconnected so that each sounds even if only one is triggered. This is a definite advantage in a house with several floors, as a detector on one floor might not be heard by occupants on another.

Ideally, each room and hallway in a house should have at least one smoke detector. As a minimum, there should be a detector between each bedroom and the rest of the house and a detector at the head of each stairway. Other important locations include utility rooms, the basement, the attic, and closets containing heating equipment.

Installing battery-powered detectors is merely a matter of attaching them to a wall or ceiling with screws. Permanently wired detectors should be installed by an electrician. Mounting instructions come with detectors. All types should be located as high in a room as possible, away from dead airspaces in corners and at the ends of hallways, and away from drafts at windows, doors, and vents. If attached to a ceiling, a detector should be at least six inches from the nearest wall; if attached to a wall, a detector should be between six and twelve inches below the ceiling. Do not install detectors where temperatures routinely fall below 40 degrees Fahrenheit or rise above 100 degrees Fahrenheit, or where normal kitchen smoke or automobile exhaust can set them off.

Test smoke detectors once a month by pressing the test button located on the front of the device. If there is no test button, hold a lit candle about six inches below the detector; if the alarm does not sound, blow the candle out and let the smoke drift into the unit. If the detector still does not sound, check the power source. Vacuum smoke detectors to keep the slots on the front free of dust. Never apply paint over smoke detectors; paint will clog the slots and prevent smoke from entering. Remove dust from inside detectors once or twice a year, or whenever you replace batteries. If false alarms become a nuisance, change the detector's location or replace it with a newer model. In any case, plan on replacing smoke detectors every ten years.

Selecting and Using Fire Extinguishers

Every home should have at least one fire extinguisher, located in the kitchen. Better still is to install fire extinguishers on each level of a house and in each potentially hazardous area, including (besides the kitchen) the garage, furnace room, and workshop.

Choose fire extinguishers by their size, class, and rating. "Size" refers to the weight of the fire-fighting chemical, or charge, a fire extinguisher contains, and usually is about half the weight of the fire extinguisher itself. For ordinary residential use, extinguishers two and a half to five pounds in size usually are adequate; these weigh five to ten pounds.

"Class" refers to the types of fires an extinguisher can put out. Class A extinguishers are for use only on ordinary combustible materials such as wood, paper, and cloth. Generally, their charge consists of carbonated water, which is inexpensive and adequate for the task but quite dangerous if used against grease fires (the pressurized

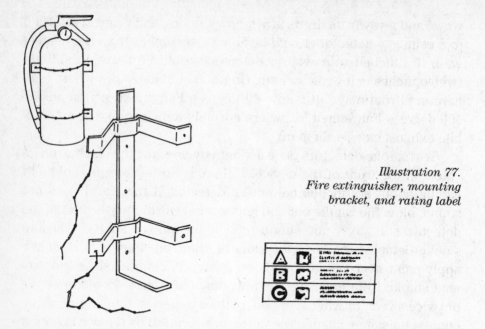

Illustration 77.
Fire extinguisher, mounting
bracket, and rating label

water can spread the burning grease) and electrical fires (the water stream and wetted surfaces can become electrified, delivering a possibly fatal shock). Class B extinguishers are for use on flammable liquids, including grease, oil, gasoline, and other chemicals. Usually their charge consists of powdered sodium bicarbonate (baking soda).

Class C extinguishers are for electrical fires. Most contain dry ammonium phosphate. Some Class C extinguishers contain halon gas, but these are no longer manufactured for residential use because of halon's adverse effect on the earth's ozone layer. Halon extinguishers are recommended for use around expensive electronic gear such as computers and televisions; the gas blankets the fire, suffocating it, and then evaporates without leaving chemical residue that can ruin the equipment. Another advantage of halon is that it expands into hard-to-reach areas and around obstructions, quenching fire in places other extinguishers cannot touch.

Many fire extinguishers contain chemicals for putting out combination fires; in fact, extinguishers classed B:C and even A:B:C are more widely available for home use than extinguishers designed only for individual types of fires. All-purpose A:B:C extinguishers usually are the best choice for any household location; however, B:C

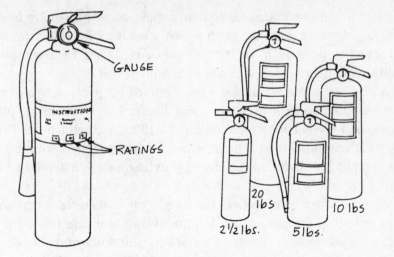

Illustration 77A. Fire extinguishers come in several sizes

extinguishers put out grease fires more effectively (their charge of sodium bicarbonate reacts with fats and cooking oil to form a wet foam that smothers the fire) and so should be the first choice in a kitchen.

"Rating" is a measurement of a fire extinguisher's effectiveness on a given type of fire. The higher the rating, the more effective the extinguisher is against the class of fire to which the rating is assigned. Actually, the rating system is a bit more complicated: rating numbers assigned to a Class A extinguisher indicate the approximate gallons of water needed to match the extinguisher's capacity (for example, a 1A rating indicates that the extinguisher functions as well as about a gallon of water), while numbers assigned to Class B extinguishers indicate the approximate square footage of fire that can be extinguished by an average nonprofessional user. Class C extinguishers carry no ratings.

For protection on an entire floor of a house, buy a relatively large extinguisher; for example, a model rated 3A:40B:C. These weigh about ten pounds and cost around $50. In a kitchen, choose a 5B:C unit; these weigh about three pounds and cost around $15. For increased kitchen protection, it is probably better to buy two small extinguishers than a single larger model. Kitchen fires usually start small and are easily handled by a small extinguisher; smaller extinguishers are more manageable than larger ones, especially in confined spaces; and, because even a partly used extinguisher must be recharged to prepare it for further use or replaced, having multiple small extinguishers makes better economic sense.

A 5B:C extinguisher is also a good choice for protecting a garage, where grease and oil fires are most likely. For workshops, utility rooms, and similar locations, obtain 1A:10B:C extinguishers. These, too, weigh about three pounds (some weigh up to five pounds) and cost around $15. In all cases, buy only extinguishers listed by Underwriters Laboratories.

Mount fire extinguishers in plain sight on walls near doorways or other potential escape routes. Use mounting brackets made for the purpose; these attach with long screws to wall studs and allow extinguishers to be instantly removed. Instead of the plastic brackets that come with many fire extinguishers, consider the sturdier marine brackets approved by the U.S. Coast Guard. The correct mounting height for extinguishers is between four and five feet above the floor, but mount them as high as six feet if necessary to keep them out of the reach of young children. Do not keep fire extinguishers in closets or elsewhere out of sight; in an emergency they are likely to be overlooked.

Buy fire extinguishers that have pressure gauges that enable you to check the condition of the charge at a glance. Inspect the gauge once a month; have an extinguisher recharged where you bought it or through your local fire department whenever the gauge indicates it has lost pressure or after it has been used, even if only for a few seconds. Fire extinguishers that cannot be recharged or have outlasted their rated life span, which is printed on the label, must be replaced. In no case should you keep a fire extinguisher longer than ten years, regardless of the manufacturer's claims. Unfortunately, recharging a smaller extinguisher often costs nearly as much as replacing it and may not restore the extinguisher to its original condition. Wasteful as it seems, it is usually better to replace most residential fire extinguishers rather than have them recharged. To do this, discharge the extinguisher (the contents are nontoxic) into a paper or plastic bag, and then discard both the bag and the extinguisher in the trash. Aluminum extinguisher cylinders can be recycled.

Everyone in the household except young children should practice using a fire extinguisher to learn the technique in case a fire breaks out. A good way to do this is to spread a large sheet of plastic on the ground and use it as a test area (the contents of most extinguishers will kill grass and stain pavement). To operate a fire extinguisher properly, stand or kneel six to ten feet from the fire with your back to the nearest exit. (If you cannot get within six feet of a fire because of smoke or intense heat, do not try to extinguish it; evacuate the house and call the fire department.) Holding the extinguisher upright, pull the locking pin from the handle and aim the nozzle at the base of the flames. Then squeeze the handle and extinguish the fire by sweeping the nozzle from side to side to blanket the fire with retardant until the flames go out. Watch for flames to rekindle, and be prepared to spray again.

Chimney Fire Extinguishers

If you operate a fireplace or wood-burning stove, keep on hand two or three oxygen-starving sticks, available at fireplace and wood-stove dealers. In case of a chimney fire, tossing the sticks into the flames will quickly quench a fire inside the chimney flue or stovepipe. Evacuate the house and call the fire department immediately in any case.

Planning Emergency Escape Routes

Having an emergency escape plan is vital. Although many fires can be extinguished if they are discovered early, flames can quickly get out of control. Every room in a house should contain at least two exits (window or door) in case one is blocked by fire. Keeping escape routes clear—and, if necessary, remodeling to insure ready access to the outside—can be the most important housework and home improvement you will ever do.

To plan and insure effective escape from your house, start by drawing on paper a floor plan of every level. Include all doors and windows, and every nonmovable obstacle.

Next, determine at least two ways out of every room; for example, a window to the outside and a doorway to a hall.

Third, follow each route and move or get rid of anything that could block a quick exit. Remember that during a real fire, darkness,

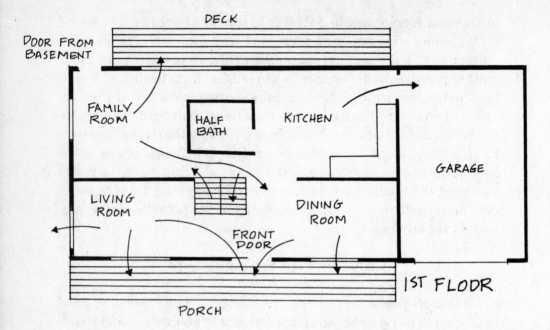

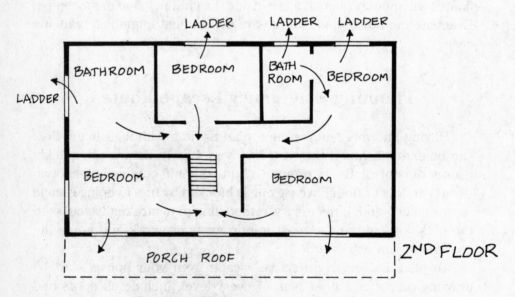

Illustration 78. Sample fire escape routes

heat, smoke, toxic gases, and confusion prevent good vision and clear thinking. Relocate large furniture blocking hallways. Free stuck windows and replace windows that are too small to climb through. Outward-opening casement windows provide nearly twice the exit space as double-hung windows of the same size.

Also replace fixed storm sashes—the old-fashioned kind that are attached from outside—with combination units that can be opened from indoors, and keep them in operating condition. Aluminum storm sashes require periodic lubrication to remain easy to operate. If their catches jam or break, replace them promptly. Hang a key in plain sight near any door with a double-cylinder lock or dead bolt. These locks offer protection from intruders breaking a hole and reaching inside, but because they can trap occupants inside, many fire codes prohibit their use.

For rooms on upper floors, try to plan escape routes that lead to a flat-roofed porch or other area on which occupants could wait to be rescued. If this is not possible, buy collapsible fire escape ladders that can be kept in easily opened storage boxes near windows. Such ladders must be sturdy enough to support the heaviest family member.

Practice evacuating the house often, even if it means climbing out of windows and shepherding small children to safety. Training, too, is a key to successful escape. During drills, practice crawling on hands and knees to exits to avoid smoke and heat. Instruct children that they cannot hide from a fire by entering a closet or by crawling under a bed. Remember that once you're outside, you should never return to a burning building.

Protecting Your Home from Lightning

If you live in an area where lightning has been known to strike, equipping your home with a lightning protection system is worth considering. Although lightning strikes have a reputation for being rare and unpredictable, they have been estimated by the National Oceanic and Atmospheric Administration to account for half as many deaths in the United States as tornadoes, hurricanes, and floods combined. According to the Lightning Protection Institute, a nonprofit or-

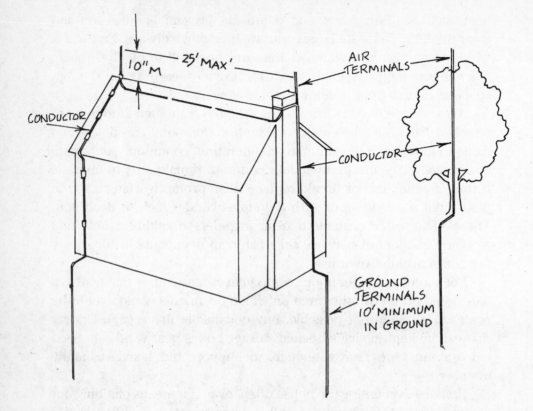

Illustration 79. Lightning protection

ganization, about eighteen thousand American homes are struck by lightning every summer.

Installing a lightning protection system is not a job for amateurs, or even for professional electricians without special training. The concept of a lightning protection system is to provide a means by which the electrical discharges associated with lightning—often in the range of thirty million volts and twenty-five thousand amps—can travel between the earth and the atmosphere without passing through and damaging nonconducting parts of a structure such as those made of wood, concrete, or masonry. Incorrect work can render the system worthless or increase the risk of fire damage. In fact, manufacturers of lightning protection equipment sell only to certified installers. To locate an approved installer, contact your local fire department or state fire marshal, or write to Underwriters Laboratories

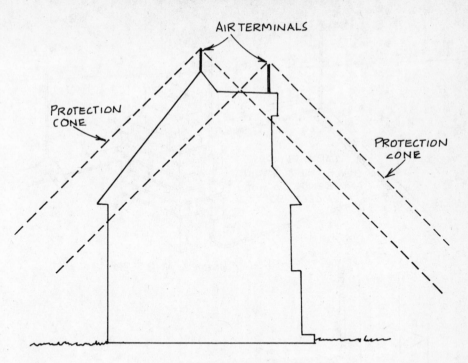

Illustration 79A. Zones of protection

Master Label Lightning Protection Program, 333 Pfingsten Road, Northbrook, IL 60062, or to the Lightning Protection Institute, 3365 North Arlington Heights Road, Arlington Heights, IL 60004. Any installer you use should observe the standards of the Lightning Protection Code compiled by the National Fire Protection Association (authors of the National Electric Code, used as the basis of most local electric codes). The average cost of installing a complete lightning protection system is between $1,000 and $2,000 for a house with uncomplicated roof geometry.

What should a system contain? Essentially, three parts are needed: air terminals, ground terminals, and conductors (wires) linking the two. Air terminals are lightning rods. They are at least ten inches long; are made of copper, copper alloy, or aluminum; and must be mounted on the highest parts of the house. On an ordinary pitched roof, air terminals are located at intervals of twenty to twenty-five feet along the ridge; the outermost terminals must be within two feet of the roof's edge. Each terminal provides a forty-five-degree conical

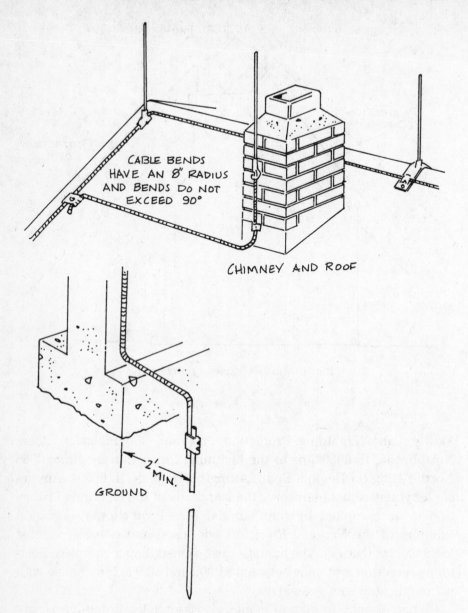

CABLE BENDS
HAVE AN 8" RADIUS
AND BENDS DO NOT
EXCEED 90°

CHIMNEY AND ROOF

2'
MIN.

GROUND

Illustration 79B. Lightning protection details

zone of protection; additional terminals are needed on the chimney and on certain house features (such as dormers) that project beyond the zones.

Conductors usually are braided copper or aluminum cables. They carry the charge of lightning from an air terminal to the ground terminals, which are rods or plates buried in the earth. Conductors must be run without bending more than ninety degrees or with a radius less than eight inches, otherwise lightning traveling along them can miss the curve and jump to a nearby surface instead of continuing to the ground. Metal gutters, television antennas, weather vanes, and other rooftop lightning attractors also must be linked to the system.

The ground terminals are made of the same materials as the other components. Industry standards call for at least two of them, at diagonal corners of the house or as far away from each other as possible and at least two feet from the foundation. The design, size, buried depth, and number of ground terminals depends on the shape and structure of the house and on the soil type, but copper-encased steel rods are common. These are driven at least ten feet into the ground.

Protection of trees higher than the house and less than ten feet away is recommended. The method is the same as for houses—one or more air terminals are installed high in the tree; the terminal then is linked to a conductor leading to a buried ground terminal. To prevent power surges in household wiring caused by lightning striking outdoor power lines near the house or by side flashes from nearby trees, a utility company technician can install a device called a secondary surge arrestor on the power lines leading to the house or at the service panel (fuse box). When activated, surge arrestors drain excess electricity harmlessly to the ground, then reopen to allow normal current flow.

Little maintenance of a lightning protection system is necessary, but owners should inspect the components periodically to make sure all of them remain connected and undamaged. Especially important are the connections of the conductors to the ground terminals, because conductors can corrode at or just below ground level. By the way, if you live in an old house equipped with ornate, old-fashioned lightning rods, it is seldom necessary to replace them with new ones—provided that they, too, remain connected to their conductors and ground terminals. However, old systems should be inspected by a certified lightning installer to determine whether they comply with current standards; in some cases conductors and ground terminals may have been inadequately sized or incorrectly installed.

Upgrading Old Wiring

ANY HOUSE CAN CONTAIN DAMAGED OR INCORRECTLY IN-
stalled wiring that can be hazardous. Older houses generally present
the greatest risks because their wiring also may be deteriorated, out-
moded, and inadequate to serve modern electrical needs. Problem
wiring generally is found in homes built before the mid-1960s; but
some wiring installed as late as the mid-1970s can be inherently unsafe, and even many installations dating from the 1980s can be made
safer by newer and improved methods and equipment.

This chapter is intended to be a guide for determining the condition of your household wiring; it also explains how a household electrical system is designed to work. Do not attempt electrical repairs
yourself without consulting additional up-to-date reference books or
other instructional materials describing repair techniques in detail,
as it is beyond the scope of this book to address them.

Also check with your local building inspector's office to make
sure amateurs in your area are allowed to perform household electrical repairs; ask about any special regulations that apply. Typically,
few restrictions prevent homeowners in rural and suburban areas
from tackling electrical repairs—generally, the only requirement is
that finished work be examined and approved by a licensed electrical
inspector, as is work performed by professionals. But in cities and for
attached housing many codes permit only professional electricians to
perform repairs. Sometimes, too, lending institutions and homeowners' insurance companies require professional repairs even if
local ordinances do not.

How Electricity Works

One way of describing electricity is to compare its behavior to water. *Voltage* (volts) is the electrical equivalent of water pressure, similar to pounds per square inch, and *amperage* (amps) is electricity's rate of flow, similar to gallons per hour. Electrical *current* consists of both volts and amps. The two multiplied together equal *wattage* (watts), which is a measure of power consumed by a *load*—for example, a lamp, appliance, or motor. Electricity itself, the water in the analogy, consists of electrons. A one-amp current flows at the rate of 6.28 billion billion—6,280,000,000,000,000,000—electrons per second.

In a plumbing system, water flows from a reservoir or pumping station and eventually returns to it, either directly through a network of sewers or indirectly via seepage through the ground itself. Electricity also follows a circular path, called a *circuit;* and as with water, the path can be direct, through a system of wires, or indirect, via the earth, which functions as a vast electrical reservoir. But there is a difference: unlike water, which will flow away from its source any time gravity draws it or artificial pressure pushes it, electricity will only flow when both routes, away from and back to its source, are un-

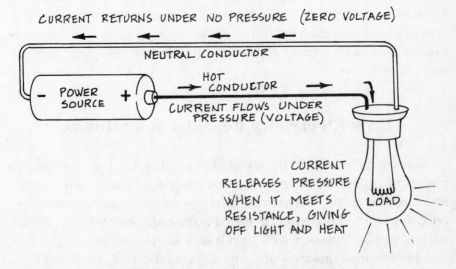

Illustration 80. How electricity works

235

blocked. When outgoing electricity returns to its source by entering the earth the condition is called *grounding*. Electricity entering the earth behaves like water pouring into a reservoir; its voltage dissipates and the electrons comprising the charge disperse, eliminating the current.

Another important aspect of electricity is that the path it follows in making a circuit always is the one offering the least *resistance*, or blockage. In this way, too, it is like water. Resistance slows the flow of electrons and converts voltage energy to heat, diminishing the current. Different materials through which electricity attempts to pass offer differing degrees of resistance; those that bar electricity are called *insulators*, and those that allow it to flow are called *conductors*. The larger a conductor's diameter the easier it is for electrons to pass; therefore, less heat from resistance is generated and less voltage is lost. On the other hand, if a conductor is too small or receives a disproportionately large current, the resistance heat generated can be sufficient to destroy the conductor and start a fire.

Distance also affects resistance. The farther current travels along a given conductor, the more resistance it encounters. Again as with water, more pressure, or voltage, is needed to propel electricity over long distances than over shorter ones. Therefore, given the choice of two equally conductive paths, a current always will follow the shorter one because it is the path of least resistance.

Wires that carry electricity are, in effect, conductors, and most electricians refer to them by that name. Conductors supplying current to a load are called *hot*; those returning it to the source are called *neutral*—or, if they carry electricity directly into the earth, *grounding*—conductors.

How Electricity Reaches Your House

Electricity generated by power plants is delivered to customers in cycles, or pulses, of *alternating current* (AC). In the United States and Canada, pulses arrive at the rate of one every sixtieth of a second; therefore, the current is called sixty-cycle, or 60-Hertz. Unlike current from a battery, which travels at uniform voltage and in a continuous stream—this type of current is called *direct current* (DC)—alternating current travels in waves whose voltage constantly varies

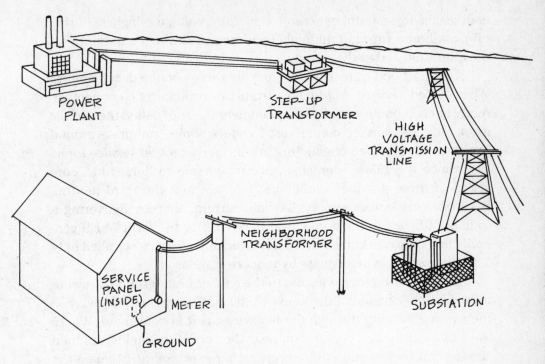

Illustration 81. How electricity reaches your home

between zero and whatever maximum amount results in a cycle's delivering the desired average voltage. AC is less expensive to produce than DC and can be more easily sent over long distances because its voltage can be increased or reduced by installations called *transformers*.

Typically, current from power plants is generated at around 15,000 volts. For long-distance transmission, step-up transformers raise power plant voltage to as much as 765,000 volts; regional substation transformers then reduce this for local distribution to between 4,000 and 12,000 volts. Power leaving regional substations travels to neighborhood transformers, the familiar large metal cylinders on utility poles. These step down power further to an average of 120 volts, which is the amount carried by most household circuits.

Alternating 120-volt current rises to as much as 170 volts during each cycle. In addition, average voltage delivered by local power companies can vary by as much as 10 percent (the exact amount usually is regulated by law) due to changing customer loads over the

237

course of a day, month, or season. Generally, voltage is highest during off-peak hours (around midnight) and lowest during peak hours (in the afternoon and early evening).

Neighborhood transformers typically serve about a dozen homes. Wires called *service conductors* from the transformer connect directly to each home, usually at a point near the roof called the *service drop*, although service conductors to some homes run underground (underground service conductors are properly termed *service laterals*). Since the 1950s, municipal electric service to homes has consisted of three service conductors: two hot and the third neutral. Both hot conductors supply 120-volt current, thereby delivering a total of 240 volts to the home. Homes built prior to World War II generally had only one hot service conductor; such service supplied only 120 volts, which is inadequate by modern standards.

Service conductors, whether to the roof or underground, are connected on the outside of the house to the *electric meter*, which reads the wattage flowing through the hot wires as it is consumed. Where the service drop is a connection near the roof, the conductors then rise several inches and enter a vertical pipe, or *conduit* (called the *service head* or *service drop*), that leads to the meter. The reason the service conductors rise to the service head is to form a loop that prevents rainwater from running along the conductors into the conduit and beyond. Power company responsibility for the repair and safety of wiring ends at the service drop, not the service head, but the company also owns and controls access to the electric meter.

From the meter, the lines run horizontally inside another length of conduit through the wall to the *service panel*, which usually is a gray metal box mounted on a wall in the basement, utility room, or garage. In older houses, service panels sometimes are located in closets. This is dangerous; panels should be located in open, dry locations with three feet of unobstructed space in front and plenty of light for working. Check the label inside the panel door. (For instructions on opening a service panel safely, see page 250.) Type 1 panels are for interior, dry locations only. Some electrical codes permit exterior-mounted panels; these should be labeled Type 3 or Type 3R.

Electricity Inside Your House

All three service conductors enter the panel at the top. There, the neutral conductor terminates at a connection called the *neutral bus bar*, while the two hot conductors end at connections with the terminals of the *main disconnect*. The latter is either a linked pair of *circuit breakers* or a pair of large *fuses*. Tripping the breakers or pulling the fuses prevents electricity from flowing beyond them; doing so thus shuts off all power to the house (circuit breakers and fuses are described in detail farther on). The neutral bus bar is also connected to the *grounding electrode conductor,* a copper wire linked with the earth via a buried metal rod or water supply pipe. Its function is to

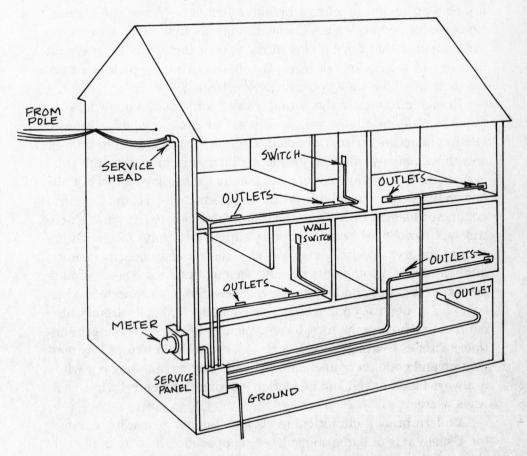

Illustration 82. How electricity enters a home

239

give current supplied by the service conductors a secondary return path—through the earth—and thereby guarantee that the house, to which the service panel and the neutral service conductor are attached (the conductor is attached to the panel), is always at the same voltage with respect to the earth nearby. The amount of this voltage is virtually zero, as any returning current encountering more resistance as it travels along the neutral conductor than is offered by the grounding conductor will follow the grounding conductor in preference to the neutral and dissipate into the earth.

Two metal plates called *hot bus bars* are attached to the base of the disconnect; these plates distribute current from each of the hot service conductors to the household *branch circuits* that serve the receptacles (wall outlets), fixtures, and appliances throughout the home. The connections of the branch circuits to the hot bus bars are also protected by circuit breakers or fuses. (All of the connections mentioned are located beneath a protective metal plate in the service panel that covers everything except the switches of circuit breakers or the sockets of fuses. Before removing the plate to examine them, read the safety precautions on page 250.)

Branch circuits are the wiring with which most homeowners are familiar. Modern circuit wiring consists of *cable*, individually insulated conductors (wires) bundled together inside a flexible tube or sheathing, usually of tough plastic. Ordinary circuits carry 120 volts and are made up of at least one hot conductor leading from a hot bus bar on the service panel to a group of outlets or loads, and a neutral conductor leading back from the outlets or loads to the panel's neutral bus bar. Some branch circuits supply 240 volts for powering energy-hungry appliances such as electric baseboard heaters, kitchen ranges, clothes dryers, and air conditioners. These circuits consist of two hot conductors, each connected to a separate hot bus bar, plus a neutral conductor. (Occasionally, 120-volt circuits also contain two hot conductors; these typically are used to connect multiple switches to a single load, such as a ceiling light fixture.) By convention and code recommendation, insulation surrounding hot wires is always black or red and insulation surrounding neutral wires is always white.

Modern branch circuit cable also contains a grounding conductor. Usually it is of bare (uninsulated) copper like the wire in the service panel; and it, too, provides an alternative path for current returning to its source or to the earth. However, the grounding conductor in a branch circuit cable is designed purely as a safety feature; it becomes a path only when current is accidentally diverted from the neutral conductor by a short circuit or a ground fault (both of these hazards are described below) and normally carries no current. Like neutral conductors, branch circuit grounding conductors are connected to each receptacle or load in a circuit and terminate at the service panel's neutral bus bar (which, as was mentioned, is connected to the earth). Some panels contain a separate *grounding bus bar;* but this, too, is linked to the same grounding electrode conductor as the neutral bus bar.

Ensuring Safety

Four features of modern household electrical systems insure safety from fire and electric shock. These are *overcurrent protectors* (fuses and circuit breakers), *continuous grounding, polarized receptacles,* and *ground-fault circuit interrupters* (GFCIs).

Overcurrent Protectors

Overcurrent protectors interrupt electrical circuits that receive too much current before they can develop dangerously high heat from resistance and cause a fire. An overcurrent can occur when too many loads are attached to the circuit, thereby excessively increasing the demand for power; when too-powerful a charge is sent through a circuit due to lightning striking service conductors or other power-company wiring; and when a broken connection or faulty insulation on circuit conductors permits electricity to flow through a complete circuit without passing through a load.

This last condition is called a *short circuit,* and usually is the result of contact between hot and neutral conductors on the same circuit. A common cause is frayed insulation that allows the conductors to touch. When a short circuit is produced by contact between a hot conductor and another conductive surface leading to the earth, the condition is properly called a *ground fault.* Both are hazards—not only because they can produce fire but because the conductive sur-

241

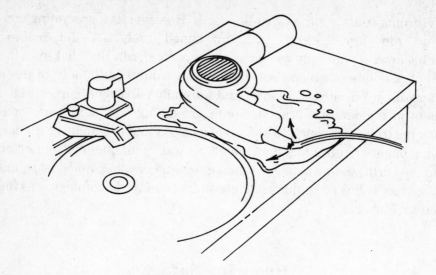

Illustration 83. A ground-fault scenario

faces in contact with the straying current become electrically charged. They are, in fact, part of a new circuit that includes them; subsequently, anyone or anything touching them can become part of the circuit also. For a person or an animal, such contact can result in severe, possibly fatal, electric shock.

Fuses are an early variety of overcurrent protector. Essentially, a fuse is simply a strip of heat-sensitive metal surrounded by a protective insulator and installed along the hot conductor of a circuit so that current on its way to a load flows through it. Should more heat develop along the conductor than the strip is designed to withstand, the strip melts, breaking the conductor and ending the current flow.

Circuit breakers are a more recent invention. They accomplish the same thing as fuses but do so by triggering a spring that separates the electrical connection to the circuit when heat is excessive. While a blown (melted) fuse must be replaced to restore current to the circuit, a circuit breaker need only be reset.

Fuses and circuit breakers are rated according to the amperage they allow to pass, and these ratings are printed or stamped on them where they are plainly visible. Fuses or breakers controlling a main disconnect typically are rated at 100 to 200 amps, depending on the number of branch circuits the service panel can accommodate. Those controlling the branch circuits usually are rated for 15 or 20

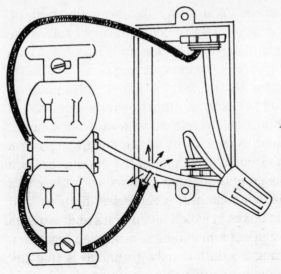

Illustration 83A. Short circuit

amps if the circuit carries 120 volts, and for 30 amps or more if the circuit carries 240 volts.

Never install a fuse or circuit breaker having a higher rating than that of the circuit cable it is designed to protect. Obviously, doing so allows excessive current to flow that can overheat the wiring. An electrician can examine the service panel connections and tell you whether any fuses or circuit breakers presently installed are inappropriate. For your part, never replace any of these devices with new ones having higher ratings, even if—*especially* if—they fail frequently. Instead, if a circuit fails more than once, consider the situation a warning and call an electrician to determine and remedy the cause.

Continuous Grounding

Much of the continuous grounding system in modern household wiring is described previously (see the section "Electricity Inside Your House"). Such systems have been incorporated in electrical codes only since the early 1960s; their purpose is to provide an emergency path through which electricity can travel safely to the ground—via the neutral bus bar—from practically any point on a household circuit should a fault occur that diverts electricity from the circuit's normal conductors. Presumably, the grounding path offers less resistance than any other path the current might follow, including one through the body of a person accidentally in contact with it. (In practice, when a short circuit occurs, causing current to travel along a grounding conductor to the service panel, the condition blows a fuse or trips a circuit breaker, shutting down the circuit.)

In a modern grounded household electrical system, branch circuit grounding wires are attached to every part of the electrical system that could result in a short circuit should one of the conductors in a circuit's cable break or loosen and come in contact with the part's metal surface. This includes all switches, receptacles, fixtures, and junction or outlet boxes (the boxes in which receptacles and switches are housed inside walls or that contain conductors spliced together).

One sign that your wiring is continuously grounded is that the wall receptacles feature sockets for three-prong plugs. Another, of course, is the presence of branch circuit grounding wires connected either to the neutral bus bar or to the grounding bus bar inside the service panel.

However, if some receptacles are of the three-slot variety and others are two-slot models, which are older, do not assume that the three-slot receptacles are continuously grounded. It is common for old-fashioned two-slot receptacles to have been replaced with three-slot models for a variety of reasons—virtually all mistaken—even though the branch circuit wiring may have no grounding conductor. Because the presence of three-slot receptacles can give the misleading impression that the electrical system is continuously grounded, installing them in circuits that are not grounded is prohibited by the National Electrical Code, on which virtually all local codes are based (for more on the National Electrical Code, see page 249).

You can determine whether a three-slot receptacle is connected to a continuous grounding system by testing it with a *voltage tester* (an inexpensive device consisting of a small bulb with two wire probes, available at hardware stores) and by inspecting the connections. Do not use an outlet analyzer, which resembles a 3-prong electrical plug having three indicator lights. These cannot detect certain serious grounding flaws, especially so-called bootleg grounds, described on p. 264. (For instructions on how to test receptacles and other electrical items, and to learn how to safely inspect service panel connections, see the section "Inspecting and Evaluating House-hold Wiring," page 248.)

Polarized Receptacles

Polarized receptacles have sockets with vertical slots of two different lengths. Correctly wired, the shorter slot is linked to the circuit's hot conductor and the longer slot is linked to the neutral conductor. Appliance cord plugs having prongs of different lengths can fit such receptacles in only one position. Because such cords are wired so the hot conductor is attached to the smaller prong and the neutral conductor is attached to the larger prong, the flow of current to and from the appliance is consistent with the flow of current along the branch circuit wiring. That is, the prongs insure that the plug can enter the receptacle in only one position; thus, they prevent an appliance from being plugged in "backwards," with current reaching the appliance through the neutral conductor instead of its hot counterpart. *Polarity* is the term describing the separateness of the hot and neutral conductors; receptacles that maintain this condition are called *polarized*.

What is the hazard of reversing the hot and neutral conductors in an appliance? None, from an electrical standpoint; the appliance will operate either way. But conductors intended to carry current generally are buried deeper inside an appliance and are less likely to jar loose, suffer wear, or be touched than neutral conductors.

Wiring in an ordinary floor or desk lamp is an example. Typically, the hot conductor fastens to a well-protected terminal located far inside the lamp socket where there is little likelihood that the connection will be disturbed. However, the neutral conductor connects close to the socket's outer surface and is kept from touching it only by a thin cylinder of cardboard acting as insulation. If the cardboard becomes worn or brittle, twisting the socket to change a lightbulb can bring the neutral terminal into contact with the metal outer covering of the socket. Ordinarily, this would present little danger because neutral conductors carry virtually no current as long as they are connected to the rest of the circuit. But if the neutral is hot instead, and you are holding the lamp when the contact occurs, you can—and probably will—receive a shock.

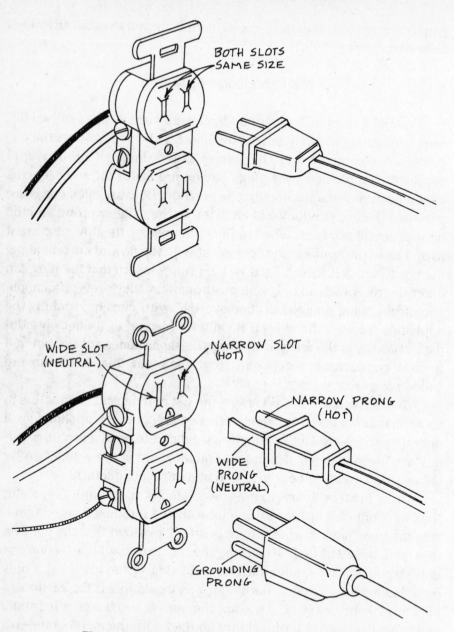

BOTH SLOTS
SAME SIZE

WIDE SLOT
(NEUTRAL)

NARROW SLOT
(HOT)

NARROW PRONG
(HOT)

WIDE
PRONG
(NEUTRAL)

GROUNDING
PRONG

Illustration 84. Nonpolarized and polarized plugs

Most polarized receptacles also have D-shaped third slots accompanying each pair of verticals. Three-slot receptacles are designed to accept the grounding lugs of three-prong appliance plugs, which are found on appliances containing a grounding wire. This arrangement links the appliance with the grounding system of the household circuit, thereby extending the protection of continuous grounding beyond the receptacle to the appliance itself.

Ground-Fault Circuit Interrupters

Fuses, circuit breakers, and continuous grounding are designed primarily to protect household wiring from fire, not people or animals from electric shock. As a result, sometimes a ground fault can occur that is so slight it fails to blow a fuse or trip a breaker, yet it allows sufficient escaping current to cause a shock.

A typical example might be a damaged power cord that allows contact between the hot conductor of a hair dryer and the damp surface of a bathroom counter. Water has excellent conductive properties and can cause electricity to "leak" from the conductor into it whether or not the appliance is turned on, even if the moisture is only a film and the cord is barely cracked. Because much of the electricity in the conductor remains contained, the fuse or circuit breaker at

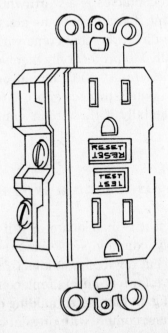
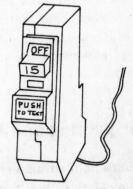

Illustration 85. Ground-fault circuit interrupters.

the service panel likely will remain intact. But should you touch the damp counter while at the same time touching a grounded surface such as a nearby metal faucet, electricity will travel from the conductor, through the moisture, *through you*, through the faucet, and along the plumbing to the earth. As little as six thousandths of an amp (.006 amps) can overwhelm the impulses governing the heart; therefore even a tiny fraction of amperage diverted from a standard 15- or 20-amp household circuit can be fatal.

Ground-fault circuit interrupters (GFCIs) are special receptacles designed to act like circuit breakers and eliminate the shock hazard from ground faults. A GFCI protects a receptacle or circuit by monitoring the current passing through it and shutting it down within one fortieth of a second if an imbalance—signaling diverted current—is detected. Not coincidentally, the amount of current imbalance needed to trigger a GFCI is only five thousandths of an amp (.005 amps), less than the amount that can disrupt a heartbeat.

The latest versions of the National Electrical Code recommend GFCI receptacles for new wiring at all locations within reach of sinks, tubs, and other water-supply outlets, and in all rooms and areas of buildings that can be damp, including basements, laundry rooms, and outside. Installing GFCI receptacles is a worthwhile remodeling activity, and not a difficult one if you have some electrical experience and can follow the written directions that come with the devices. GFCI receptacles can be installed to protect only themselves or to protect other outlets on the same circuit. (GFCI circuit breakers, which are installed at the service panel, protect entire circuits but are expensive and can trip accidentally if the circuit contains many outlets.)

Inspecting and Evaluating Household Wiring

Household wiring needs attention if it is inadequate or if it presents a safety or fire hazard. Obviously, common sense dictates repairing hazardous wiring immediately, but few communities (if any) have laws requiring it. Neither are there laws requiring replacement of decrepit or obsolete residential wiring. In fact, most building code regulations state only that *new* work must comply with present code standards; existing wiring and equipment may remain in use "as is" so long as it is not modified or added to. (However, household wiring that is not "up to code" usually lowers a home's value and may hamper selling it.)

Most electrical codes are based on the National Electrical Code (NEC), first published in 1897 and updated every three years by the National Fire Protection Agency, a private nonprofit organization. Although a copy of the latest N.E.C. is available in most libraries, the work is so large and complicated that, generally speaking, only professional electrical inspectors and building code officials can make sense of it. As an amateur considering wiring modifications, your best bet is to study up-to-date nontechnical books on household wiring and then to consult an electrician or inspector for specifics regarding your house and regulations that apply in your community.

Even if local regulations permit amateur electrical work, inspecting and rewiring are tasks usually best left to professionals. Handling old wiring can often create hazards where there were none, and an experienced inspector or electrician can make sense of old-fashioned and amateur techniques too obscure for books.

Service Conductors

If you live in a house built before the 1960s, two-wire service conductors—delivering only 120 volts—may still supply it. Obviously, such current cannot run 240-volt appliances or other items such as electric heating and cooling systems. The amount of amperage such current delivers (30 amps in many homes built prior to the 1950s, 60 amps in others) also is not adequate to power more than a few 120-volt loads at the same time. If your home is supplied by two-wire service conductors, it generally is the responsibility of the electric company to replace them with modern three-wire conductors and a new electric meter. However, as the homeowner, you will be responsible for installing everything else, including a new service panel and branch circuits.

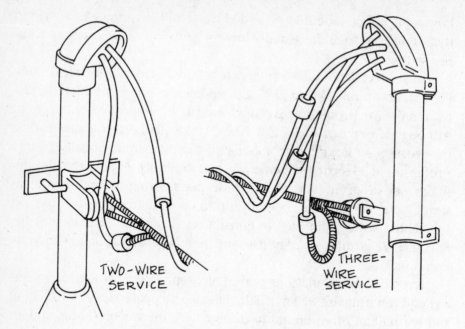

TWO-WIRE SERVICE

THREE-WIRE SERVICE

Illustration 86. Service drops: Old and new

Service Panel and Grounding Electrode

Service panel capacity currently must be at least 100 amps to comply with National Electrical Code recommendations (200-amp capacity is required for electric heating or if service conductors are more than three hundred feet in length). You can determine the capacity of a service panel by opening its hinged access door and reading the label printed on the inside, and by adding together the amperage ratings of the fuses or circuit breakers comprising the main disconnect.

To open a service panel safely, first be sure your hands are dry and that the floor in front of the panel is dry also. If in doubt, stand on a thick piece of dry lumber or plywood. Examine the outside of the panel for signs of moisture—especially fresh rust—and blackening from sparks or smoke. If you see any, call an electrician to open the panel; do not touch it yourself.

Assuming that the outside of the service panel appears dry and undamaged, open it with one hand while keeping your other hand in

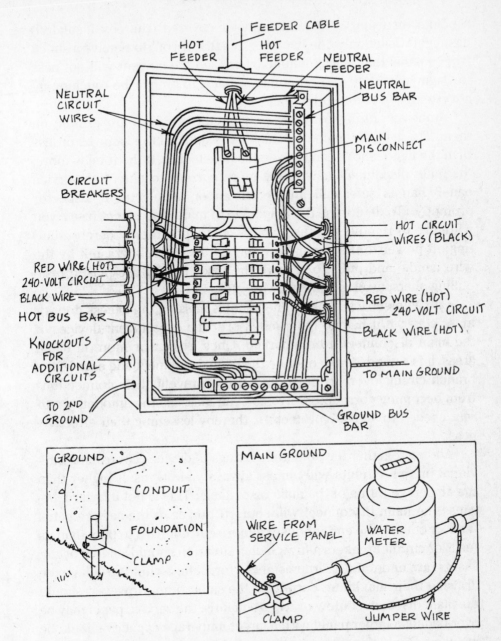

FEEDER CABLE

HOT FEEDER

HOT FEEDER

NEUTRAL FEEDER

NEUTRAL CIRCUIT WIRES

NEUTRAL BUS BAR

MAIN DISCONNECT

CIRCUIT BREAKERS

HOT CIRCUIT WIRES (BLACK)

RED WIRE (HOT) 240-VOLT CIRCUIT

BLACK WIRE

HOT BUS BAR

RED WIRE (HOT) 240-VOLT CIRCUIT

BLACK WIRE (HOT)

KNOCKOUTS FOR ADDITIONAL CIRCUITS

TO MAIN GROUND

TO 2ND GROUND

GROUND BUS BAR

GROUND

CONDUIT

FOUNDATION

CLAMP

MAIN GROUND

WIRE FROM SERVICE PANEL

WATER METER

CLAMP

JUMPER WIRE

Illustration 87. Service panel and grounds

a pocket or using it to hold a plastic-covered (thereby insulated) flashlight. Using only one hand insures that if you do receive a shock from a panel that is accidentally energized the current will not pass through your body to a grounded surface you might be touching inadvertently with your other hand.

Amperage ratings of fuses and circuit breakers are printed on them. Those of the main disconnect should total the same as or less than the figure shown on the service panel label. If the total is more, the main disconnect may allow more current to enter the service panel than is safe; call an electrician or an electrical inspector promptly. (To determine the amperage of main disconnect fuses, you will have to remove the fuse blocks containing them, thereby shutting off power to the panel. To remove a fuse block, grasp it by the wire handle and pull. To remove the fuses inside, use plastic fuse-pulling pliers, available at hardware stores.)

Call an electrician also if the panel is missing the access door or inner cover, if it has no inner label, or if the overcurrent devices of the main disconnect are unmarked. Once a year, or more in damp areas, it is a good idea to operate the main disconnect and each of the branch circuit overcurrent protectors to prevent their connections from becoming corroded. Corrosion can increase the amount of current needed to activate protectors, thereby lessening their effectiveness.

Adding together the amperage ratings of devices controlling individual branch circuits will almost always yield a result higher than the amperage rating of the main disconnect. This is not a hazard, because the main disconnect will shut off power to the service panel (and, of course, the entire house) if more current tries to pass than its fuses or circuit breakers allow. Such a situation should seldom occur, as it is rare under most circumstances for all household circuits to be drawing their maximum current at the same time. However, should the main disconnect blow or trip frequently, the service panel may be overloaded; a larger panel with a larger amperage capacity should be installed. Books for amateurs are available with instructions describing how to do this, but most homeowners are better advised to hire a licensed electrician to perform the task.

You should also inspect a service panel for an adequate grounding electrode conductor. To do this, search first for the grounding electrode—a bare or sheathed copper wire leading from a hole in the panel to a cold water pipe nearby, or to the top of a metal rod penetrating the foundation, or to a rod driven vertically into the ground outside near the panel's indoor location. If your house is supplied with water by a well, the grounding electrode may also be attached to the metal well casing.

Many building and electrical code regulations cover the connection of the grounding electrode conductor. For example, where the connection is to a pipe, the conductor must be attached beyond the water meter so that removing the meter or attaching it with non-metallic fittings will not interrupt the conductive path. (If the conductor is attached on the indoor side of the meter, look for a copper jumper wire linking the indoor and outdoor segments of pipe by bypassing the meter.) The pipe itself must extend at least ten feet underground. Be certain a grounding electrode apparently attached to a water pipe is not, in fact, attached to a gas-supply line. As unbelievable as it may seem, such connections are not uncommon; obviously, they create a serious explosion hazard and demand immediate rewiring to a different location.

Where the connection is to a metal rod in the foundation, the rod must connect with other embedded rods at least twenty feet in length. Where the connection is to a metal rod outside, the rod must extend at least eight feet vertically into the earth (in dry regions, some codes require two rods be used) or be the end of a thick copper wire at least twenty feet long buried two and a half feet deep alongside the house.

If you cannot locate a grounding electrode conductor leading from the service panel to one of the recommended locations, or if the conductor appears loose, damaged, or inadequate (the gauge size of wire used for grounding electrodes must be six or lower; size four is recommended), call a licensed electrician or an electrical inspector for further advice and possible repair. The absence of a grounding electrode is unacceptable. Additions to the National Electrical Code even recommend supplementing electrical systems having only one of the grounding electrodes mentioned with at least one of the other types.

Inside the service panel, the grounding electrode must be securely fastened either to the neutral bus bar or to a separate grounding bus bar if the panel contains one. Checking for this connection means removing the panel's protective inner cover, which is potentially more hazardous than merely opening the access door. The wisest course is to have the inspection performed by an electrician or an electrical inspector, but if you would like to examine the inside of the panel yourself, here is the safest way to proceed.

First, follow all of the precautions described above for opening the service panel access door. Next, open the access door and shut off incoming power to the panel by removing the fuses of the main disconnect (never simply loosen fuses; always remove them completely) or by tripping its circuit breakers. Last, unscrew the bolts securing the inner cover plate and gently pull the plate away from the panel, being careful not to let it move up or down, as this might bring its surface into contact with the wiring connections behind it. *Remember: despite the fact that power to the branch circuits has been shut off by activating the main disconnect, electricity still flows into the panel via the hot service conductors. Touching these can be fatal.*

Study the wiring connections without touching them until you are completely familiar with everything you see. Hot, neutral, and grounding conductors should be uniformly color-coded and easy to identify. All conductors should be short, neatly arranged, and run along the sides of the panel, and there should be no unattached or extraneous wires. If wiring inside the panel is confusing or jumbled, suspect amateur work that could be faulty. Other signs of possible trouble are rust; brittle, cracked, or charred insulation on conductors; and evidence of sparking. If you see any of these have the service panel promptly inspected by a professional.

To replace the inner cover, take the same care as when removing it, and make sure it fits correctly before tightening the bolts securing it. When you are finished, restore power to the main disconnect.

Branch Circuit Wiring

The earliest household electrical systems still operating consist of two individual rubber-coated conductors, one hot and one neutral. The wires are run separately with at least three inches of space between them, and each is further protected from contact with other surfaces by porcelain insulators that also serve as anchors securing the wiring. Short, thick insulators to which wires are fastened are called *knobs;* their purpose is to hold wires away from framing and to

254

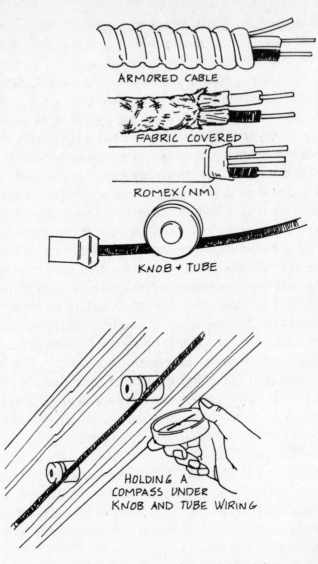

ARMORED CABLE

FABRIC COVERED

ROMEX (NM)

KNOB + TUBE

HOLDING A
COMPASS UNDER
KNOB AND TUBE WIRING

Illustration 88. Types of household wiring

provide a means of attaching them. Long, narrow insulators lining holes in which wires pass through framing are called *tubes*. Inevitably, this method of installing wiring gained the name "knob-and-tube" wiring; it was the normal residential wiring technique until shortly before World War II.

Old knob-and-tube wiring is not inherently unsafe. In fact, it is quite safe provided the wires are not overloaded by excess current

255

and the insulation surrounding each wire is intact. Unfortunately, this is seldom the case; by now most rubber insulation surrounding knob-and-tube wiring has deteriorated (as has any electrician's tape or other splicing material used to attach switches, receptacles, and other outlets), and modern houses generally use more electricity than these wires can safely carry. Of course, such wiring is rarely if ever grounded, but that in itself is not necessarily a reason for replacing it.

If you discover knob-and-tube wiring in your house, do not touch it, as doing so could damage fragile insulation or connections that for the moment are intact. You can determine whether electricity still runs through the wiring by holding an ordinary magnetic directional compass an inch or so away from each wire; if the needle moves, the wiring is "live." Naturally, if you can see that the wiring is connected to a functioning light socket or receptacle, that, too, is proof.

If a compass does not deflect when held near both wires, it is safe to assume that they have been disconnected and no longer function. If the wiring is in the basement or another area where the service panel is located, you can double-check by following it toward the panel. You should discover the ends of both wires. If you find the wires still connected to the service panel, read the safety precautions on pages 250–254 and then remove the panel's inner cover to check whether they lead to an empty fuse socket (knob-and-tube wiring is seldom, if ever, governed by circuit breakers). If they do, call a licensed electrician and have them removed.

On the other hand, if the wiring is live, locate the fuse controlling it. You can do this by turning on a light or other appliance supplied by the circuit and then removing fuses until the light goes out; or you can perform the compass test described earlier and remove fuses until you find one that causes the compass to behave normally rather than deflect. To safely remove a circuit fuse, also called a plug or Edison fuse, unscrew it by hand, being careful to touch only its nonmetal parts. Remove the fuse completely from the socket; do not merely loosen it. When replacing a circuit fuse, tighten it firmly; a loose fuse generates heat and can create a fire hazard.

Check for other lights or appliances that might have gone out. If all of the wiring to these is visible for inspection and is in good condition, there generally is no need to alter or replace it. However, it is not safe to use the wiring for purposes other than those for which it is presently used; and by all means, you should consider replacing it if remodeling is planned that will disturb or hide it. The National Electrical Code does not permit merging new wiring with knob-and-tube wiring, nor does it permit repairing defective knob-and-tube systems. If any live knob-and-tube wiring is not visible, call an electrician or an electrical inspector to appraise the situation.

Other old wiring resembles the modern style in that cable is used rather than individual wires. Again, the problem with it is deterioration. Three types of old cable may be found: black cable having a braided-fabric shell covering rubber insulation, silver-colored cable also having a braided-fabric covering, and armored cable, called BX or Greenfield (a brand name), covered with steel or aluminum ribbon wrapped in an overlapping spiral pattern.

The two fabric-covered cable varieties rot rather quickly in damp conditions and crack in dry ones (for example, attics). The silver-colored cable is especially prone to rot because it usually contains jute or paper that absorbs moisture. In addition, neither is likely to contain a grounding conductor.

Armored cable actually is still specified by some electrical codes for some uses, notably in masonry and plaster walls and in walls that may have many nails driven into them. However, the cable may not be used in damp or potentially damp locations, including airspaces in masonry walls extending below ground level. Older armored cable provides reasonable continuous grounding in that its metal sheathing will conduct errant current provided the cable is securely in contact with the outlet boxes to which the cable is attached. Modern armored cable contains a strip of copper called a *bonding strip* that is designed to carry such current internally, inside the armor instead of along it. Typically, the bonding strip on armored cable is wrapped tightly around the clamp securing the cable to the outlet box and, at the other end, to the service panel. The danger of relying on armored cable to supply continuous grounding is that circuits may not contain it in their entirety; thus neither the sheathing nor the bonding strip may establish a complete circuit for stray electricity to follow.

Fabric-covered cable is often a more likely candidate for replacement than knob-and-tube wiring, although this is a judgment call that should be made only by an experienced electrician or electrical inspector. The reason is that the conductors in cable are close enough to touch or be linked by moisture—causing a short circuit—if the insulation surrounding them is breached. The same is true of armored

257

cable; consider replacing any that appears rusty or is located in a damp area such as a basement.

Contemporary plastic-covered cable, called Romex (also a brand name) by most electricians, but officially termed NM for "nonmetallic," is virtually impervious to decay and drying unless it is used outdoors or is buried in the ground. NM cable is easily recognized by its white or cream color.

Aluminum Conductors A type of NM cable widely used during the 1960s contained potentially dangerous aluminum conductors instead of the standard copper. The reason: aluminum was—and is—cheaper. However, aluminum resists the flow of electricity more than copper and so heats up faster and expands more when electricity runs through it. Both metals shrink when current is turned off, allowing them to cool; but because the overall change in size is greater for aluminum than for copper, and because the changes occur at different rates, aluminum conductors connected to copper or brass fittings inside electrical devices such as switches and receptacles eventually loosen and corrode. Sparking resulting from the poor connections can cause a fire. According to the U. S. Consumer Product Safety Commission (CPSC), homes with aluminum wiring are 55 times more likely to be a fire hazard than homes with copper wiring.

In 1971, changes in the National Electrical Code forbade the use of electrical devices containing copper or brass terminals with aluminum wiring. Aluminum wiring continues to be available, but now there are two kinds—all-aluminum and copper-clad aluminum. If your house was built before 1965 and has had no new wiring since that time, there is little chance that aluminum wiring of either variety was used. But if your home was built between 1965 and 1971 or has had additional wiring since then, the safest course is to have an electrical inspector or an electrician examine the wiring and make a determination; identifying aluminum wiring is not as easy as it may seem, especially if it is copper-clad.

High-Resistance Faults

Deteriorated insulation on old wiring, particularly wiring covered with dust, can "leak" electrical current at a very low but still significant level. The cause simply is that the insulation no longer restricts electricity along the conductors it surrounds, and some current follows whatever minor paths of least resistance are presented by the dust, moisture, and air nearby. Because the leaking current is so slight, the overall resistance in the conductor stays virtually the same, hence the term "high-resistance" is used to describe the fault. The condition usually prevents fuses and circuit breakers from failing as a direct result, although they may fail more frequently. In any case, high-resistance faults can produce heat and should be considered fire hazards. They can also needlessly increase utility bills, much like a leaking faucet.

You can test for high-resistance faults by using your electric meter to determine whether power is being consumed even when no electrical loads have been placed on the circuits. The procedure is fairly simple but must be done thoroughly to avoid mistakes.

Start by removing all light bulbs and fluorescent tubes from fixtures controlled by wall switches and by unplugging all appliance cords from receptacles. To avoid overlooking anything, operate each switch to determine which position supplies power; turn the switch off to remove the bulb, tube, or plug; then return the switch to the on position and secure it with tape. Do not overlook obscure light fixtures in closets, attics, and garages. Also be alert for switch-controlled receptacles; these must be left on.

Shut off power at the service panel to circuits controlling permanently wired appliances such as whole-house fans, water heaters, and central heating and cooling equipment. Then disconnect the circuit wires from each of those items and cover their ends for safety with wire connectors. (If you are unsure of how to safely disconnect or reconnect permanently wired items, you may want to call an electrician to help you perform the test.) Afterward, restore the power and turn on and tape all switches to disconnected wiring.

Now for the test. Go outside and observe the circular rotor inside the electric meter for at least one minute. If the rotor remains motionless, your branch circuit wiring is in good condition. If the rotor moves, electricity is being consumed, either because of a mistake in your preparations for the test or because of a high-resistance fault somewhere in the system.

If the rotor moves, go to the service panel and shut off power to all the circuits, but do not shut off the main disconnect. Observe the meter again; if the rotor still moves, the problem is located in the service panel or in the service conductors leading to the house.

If the rotor does not move, go back and forth between the service panel and the meter, restoring individual circuits one at a time and observing their effect on the rotor. When you find a faulty circuit—one that causes the rotor to move—check to see whether any three- or four-way wall switches are installed on it; such switches permit operating a light fixture from two or more locations. If so, move each switch to its opposite position and observe the effects on the meter one at a time. Also double-check to make sure no loads remain connected to the circuit. If the rotor continues to move, consider the circuit faulty. Identify it by marking it at the panel, then shut off the power to it and continue to test the remaining circuits.

When you have finished, you may wish to inspect conductors inside electrical outlet boxes and elsewhere for dirt and damage as is described below. However, locating high-resistance faults is usually a job for an electrician; there may be several faults along a circuit, a fault may be in wires located behind a wall, or the wiring may be so badly deteriorated that the fault essentially occurs along most or all of its length.

Making Repairs

Wiring

It is best not to attempt repairing obsolete wiring if it is working, as this may damage it. (As was mentioned earlier, the National Electrical Code does not allow repairing knob-and-tube wiring in any case.) Instead, replace the wiring altogether. Besides the faults already covered, additional wiring flaws and their solutions are listed below:

- Cable attached to edges of exposed framing: not allowed by NEC. Circuit cable must pass through holes drilled at least two inches from edges in framing members.
- Cable hangs loosely along framing: anchor cable to the sides of framing with staples made for the purpose (plastic staples are preferred). Locate staples one and a quarter inch from framing

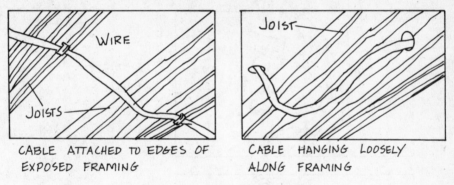

CABLE ATTACHED TO EDGES OF
EXPOSED FRAMING

CABLE HANGING LOOSELY
ALONG FRAMING

Illustration 89. Wiring faults

edges; spacing should be at least every four and a half feet and within twelve inches of each electrical box.

• Exposed splices in cable conductors: not allowed by NEC. Enclose splices inside approved junction box. (Junction boxes must remain accessible; they cannot be hidden behind walls or other surfaces.)

• Cable conductors spliced with electrical tape: replace with wire connectors.

• Unclamped cable enters metal electrical box, permitting cable to rub: disconnect conductors inside box, install cable clamp, then reconnect conductors. (Clamps are not necessary if cable enters plastic outlet box and is anchored within twelve inches of the box.)

Outlets and Switches

If your house contains obsolete branch circuit wiring you can be practically assured the receptacles, fixtures, and switches connected to it also are obsolete. But if the wiring is still sound the outlets may be, too. Have a licensed electrician or electrical inspector check wiring installed before World War II. (Just be aware that examining devices and wiring that old is risky; the process can easily damage brittle insulation and fragile connections, creating hazards where none existed.) You can inspect newer outlets yourself without much trouble, but be sure to work safely and do not proceed until you understand the procedures. Do not attempt to test high-voltage (240-

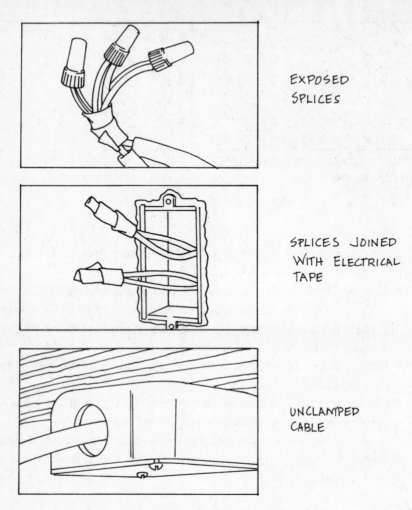

EXPOSED
SPLICES

SPLICES JOINED
WITH ELECTRICAL
TAPE

UNCLAMPED
CABLE

Illustration 89A. More wiring faults

volt) outlets—for example, those for electric ranges. Call an electrician instead.

Obviously, if a receptacle, fixture, or switch doesn't work or behaves erratically it constitutes a hazard and must be repaired or replaced promptly. Other signs that an outlet needs attention are sparking, buzzing, cracks or other damage, and evidence of arcing (usually scorch marks). Condemn also receptacles whose sockets do not grip plugs securely and switches that require delicate positioning of the buttons or toggle to make electrical contact.

Illustration 90. Testing a two-slot receptacle for grounding

For testing 120-volt outlets, use a voltage tester. To test polarized receptacles you know are correctly grounded (see p. 245), you can also use an outlet analyzer.

Outlets To test a receptacle with a voltage tester, start by inserting the probes into the two vertical slots of each socket. It is wise to observe the "one-hand" precaution described on pages 250–51. With the probes in the slots, the tester should light, indicating that power reaches the receptacle.

Next, if the receptacle is a two-slot model, insert one probe into one of either socket's slots and touch the other probe to the screw securing the receptacle's cover plate (the screw holds the plate against an outlet box inside the wall). If necessary, scrape the surface of the screw head with a knife so the probe touches metal, not paint. Repeat the test with one probe in the other slot of the same socket.

If the tester lights in one of the positions with a probe touching the cover plate screw, the receptacle is grounded. Bright light, similar to that obtained by inserting both probes into a socket's slots, indicates effective and probably intentional grounding. Dim light suggests ineffective grounding or a fault requiring further examination. A barely perceptible glow can be ignored; the current flow it signals is due to the grounding supplied by the metal outlet box to which the cover plate screw is attached.

Presumably, when the tester lights with one probe touching the cover plate screw it is because the other probe is in contact with the hot conductor of the branch circuit cable attached to the receptacle. The tester should not light or even glow with the probe in the remaining slot; if it does, this signals a serious electrical problem that must be examined by a professional.

If the receptacle is a three-slot model, perform the grounding tests by inserting one probe into the D-shaped holes instead of touching the cover plate screw. The tester should light brightly with the remaining probe inserted into the corresponding socket's shorter vertical slot but not light at all with the probe inserted into the longer slot. In a correctly wired polarized receptacle the branch circuit's hot conductor is always attached to the shorter slot and the neutral conductor is always attached to the longer slot. If the tester lights with a probe in the longer slot but not in the shorter, the receptacle is wired incorrectly, with the hot conductor attached to the longer slot.

Further testing of receptacles and all testing of fixtures and switches (described below) should be done only after shutting off the power to the outlet by removing the appropriate fuse or by tripping the appropriate circuit breaker.

With the power to the circuit shut off, unscrew the cover plate surrounding the receptacle. Grasp each of the tester's probes by its insulated wire and touch the probes to all combinations of terminal screws on the receptacle. Then touch them to all combinations of conductors and the metal outlet box enclosing the receptacle. At no time should the tester light; if it does, the power is still on.

Assuming that the tester does not light, continue by unscrewing the receptacle from the outlet box at the top and bottom and gently pulling the receptacle out from the wall. Two-slot receptacles should be attached only by black (hot) and white (neutral) conductors. Three-slot receptacles should be attached also with a green-insulated or bare copper grounding wire coming from the same cable(s) as the other conductors. A three-slot receptacle featuring only a short grounding conductor connecting it to the outlet box is not effectively grounded. In fact, this arrangement—sometimes called a bootleg ground—is quite dangerous, as it may direct full current from a faulty receptacle to the outlet box and its cover plate. A person touching the plate could be electrocuted. In addition, the electrified box is potentially a fire hazard. Look for the following other faults and correct them as prescribed:

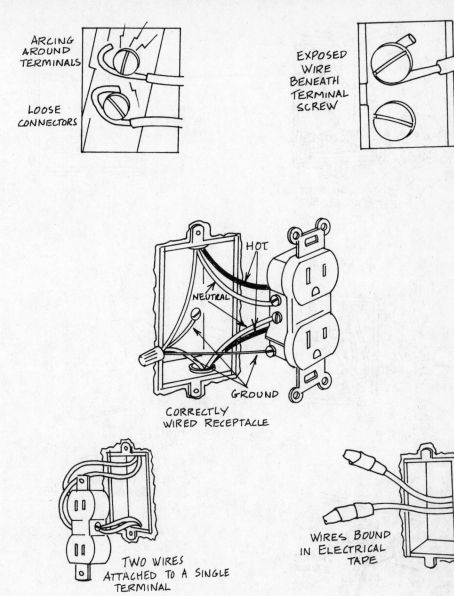

ARCING AROUND TERMINALS

LOOSE CONNECTORS

EXPOSED WIRE BENEATH TERMINAL SCREW

HOT

NEUTRAL

GROUND

CORRECTLY WIRED RECEPTACLE

TWO WIRES ATTACHED TO A SINGLE TERMINAL

WIRES BOUND IN ELECTRICAL TAPE

Illustration 91. Receptacle faults

- Loosened conductors: tighten terminal screws firmly.
- Evidence of burning or arcing at terminals (usually caused by
  loose terminal screws): disconnect conductors and polish them
  with fine sandpaper, then reconnect. Fashion new connections
  instead, if necessary. If receptacle appears damaged, replace it.

DUST AND DIRT IN
ELECTRICAL BOX

SHORT
WIRES
IN BOX

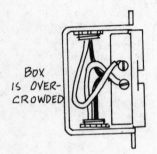

BOX
IS OVER-
CROWDED

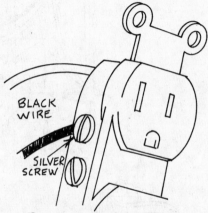

BLACK
WIRE

SILVER
SCREW

BLACK WIRE CONNECTED
TO SILVER SCREW

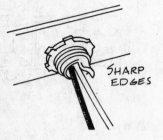

SHARP
EDGES

Illustration 91A. Receptacle faults

- Bare ends of hot or neutral conductors not completely covered
  by terminal: remove conductors from screws, trim bare wire

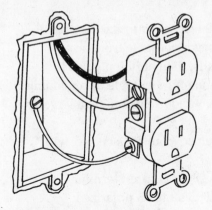

THREE SLOT RECEPTACLE
CONNECTED TO UNGROUNDED
CIRCUIT

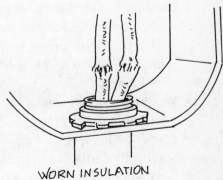

WORN INSULATION

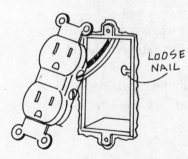

LOOSE
NAIL

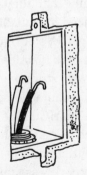

Illustration 91B.   Receptacle faults

ends shorter or fashion new connections. Reattach beneath terminal screws.

- Multiple conductors attached to the same terminal screw: remove multiple wires and join them with a plastic wire connector to a pigtail (a short length of conductor wire) the same color and gauge as the removed wires. Attach the remaining end of the pigtail to the terminal screw.
- Conductors connected with electrical tape: remove the tape and

267

join the bared ends of the conductors with a wire connector instead.

- Outlet box contains dust or other debris: blow or vacuum clean.
- Conductors too short to allow pulling receptacle from box: disconnect conductors, renew ends, then attach pigtails to each with wire connectors. Reconnect receptacle to remaining ends of pigtails. (This solution may require installing a larger outlet box.)
- Too little room in outlet box: have electrician replace box with larger model.
- Black (hot) conductors attached to silver-colored terminal screws or adjacent to longer vertical slots on polarized receptacles: reverse conductor connections so that black wires are attached to brass terminal screws next to shorter slots.
- Three-slot receptacle connected to ungrounded branch circuit cable (this is dangerous—and violates the National Electrical Code—because it misleads users into thinking such receptacles are safer than they really are): replace with two-slot receptacle, polarized if available.
- No protective sleeve on armored cable where conductors enter outlet box: install plastic protective sleeve (available at hardware stores) around conductors.
- Frayed or damaged conductor insulation: wrap damaged areas with electrical tape as a temporary fix. Have electrician replace branch circuit cable.
- Loose outlet box: reanchor securely or replace.
- Outlet box recessed into wall, preventing cover plate from fitting tightly: add extension ring to bring surface of electrical box flush with wall. Reinstall cover plate.

Switches To test switches, use a *continuity tester*, a device that supplies a small current. Start as you would for testing a receptacle, by shutting off the power at the service panel and then removing the cover plate from the wall to expose the switch. Make sure the power is off by touching the probes of a voltage tester to each terminal screw and then by touching one probe to the electrical box and the other probe in turn to each terminal screw. The tester should not light in any position.

When you are certain no power reaches the switch, pull it gently

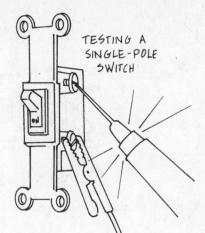

TESTING A
SINGLE-POLE
SWITCH

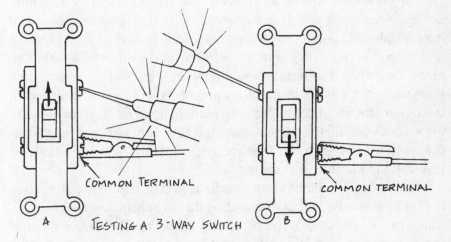

COMMON TERMINAL

COMMON TERMINAL

A   TESTING A 3-WAY SWITCH   B

Illustration 92. Testing switches with a continuity tester

from the wall. If only two conductors are fastened to the switch, indicating that it is a single-pole switch—the simplest type—disconnect them. Attach one conductor of the continuity tester to one of the switch's terminal screws and touch the other conductor to the switch's remaining screw. Operate the switch. The tester should light with the switch in one position but not the other.

If the switch has three conductors attached to it, it is a three-way switch. Label the conductors with masking tape to be certain of reconnecting them correctly, then disconnect the switch and test it

269

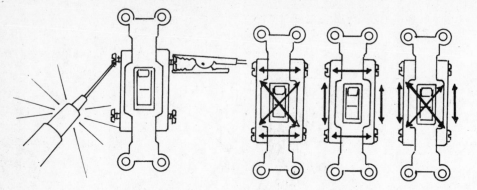

Illustration 92A. Testing a four-way switch

with a continuity tester as follows: Locate the switch's common screw terminal; usually it is colored differently from the others and may be labeled "Common." Attach one of the tester's conductors to it and touch the remaining conductor to one of the other terminals. Operate the switch; the tester should light with the switch lever in one position but not both. Without removing the conductor attached to the common terminal, touch the remaining conductor to the remaining screw. Operate the switch again; the tester should light only when the switch lever is in the position opposite that which caused the tester to light in the previous test.

A four-way switch has four conductors attached to it. After labeling the conductors and disconnecting the switch as described above, touch the conductors of the continuity tester to all possible pairs of screw terminals with the switch lever in one of its two positions. The tester should light when connected to two of the screw pairs but not the others. Flip the switch lever to the remaining position and repeat the test; again, the tester should light when connected to two of the screw pairs, but these should be the pairs that did not cause the tester to light during the previous test. In all, if the switch is good the tester will recognize four continuous paths between terminals— two for each switch position. If not, the switch is faulty and must be replaced.

Occasionally you may find a single-pole switch installed in place of one or both sockets in a receptacle. To test these switches, proceed as above to disconnect them; then, with a receptacle containing one switch, attach a conductor of the continuity tester to one of the

270

terminal screws nearest the switch and touch the other conductor to the terminal screw on the switch's opposite side. The tester should light with the switch lever in one position but not the other. Use the same technique to test a receptacle containing two switches, but test each switch separately.

Recognizing Faulty Switch Wiring Switches work by establishing or breaking the circuit on which they are installed. They will work whether they are located on the hot conductor leading to a load or on the neutral conductor returning to the source; however, for safety reasons switches must always be installed only on hot conductors. A switch installed on a neutral conductor interrupts the path to earth the neutral provides, and also permits current to reach the load and deliver a shock if the hot conductor accidentally grounds inside it.

Where only one cable enters the outlet box containing a switch, all conductors except grounding conductors should be connected to the switch, and the white conductor, if there is one, should be wrapped with black tape or painted at the end to signify that it carries live current when the switch is on. Where more than one cable enters the outlet box, its white conductors should be joined together with a wire connector, and only black, red, or deliberately marked white conductors should be fastened to the switch.

Aluminum Conductors If your house is wired with aluminum, all switches and receptacles should be marked "CO/ALR." Unmarked devices and devices marked "CU" are suitable only for copper wiring; devices marked "CU CLAD" are made for use with copper-clad aluminum wire but can also be used with copper wire. Devices marked "CU/AL" were made before the NEC changes. They were intended to be suitable for both copper and aluminum but proved otherwise; now they are allowed only for copper or copper-clad aluminum wiring.

No aluminum wiring should be installed in push-in terminals, the slots at the rear of some switches and receptacles. Aluminum wiring cannot safely be added to copper or copper-clad aluminum wire.

In all cases, switches that do not pass testing must be replaced using duplicates or updated models. When replacing or reinstalling switches, polish the ends of conductors with sandpaper before reconnecting them, or else cut them off and fashion new connections if they are nicked or burned.

271

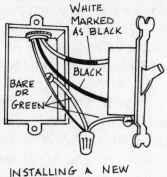

INSTALLING A NEW
SWITCH LOOP

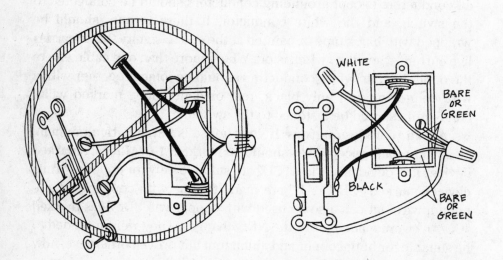

Illustration 93. Recognizing faulty switch wiring

Lighting Fixtures

Ceiling light fixtures fall into three basic categories: incandescent, fluorescent, and track. Circuit wiring for each terminates at the outlet boxes to which they are attached. As with receptacles and switches, inspection and testing is divided between examining the

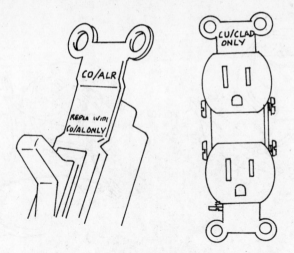

Illustration 94. Fixtures for aluminum wiring

connections between the circuit wiring and the device and examining the device itself.

Overhead and wall-mounted incandescent fixtures in old houses often pose hazards even if they apparently function normally. The reason is that homeowners typically install light bulbs having wattage ratings in excess of those for which the fixtures were designed. Overheating then occurs inside the fixture, which damages conductors and insulation and can scorch or char nearby building materials. Over the years, light fixtures also can accumulate dirt and debris that can catch fire or cause a short circuit.

To examine an ordinary ceiling- or wall-mounted fixture, proceed as you would for examining a receptacle or switch, by shutting off power to its circuit at the service panel. Operate the switch to make sure the power is off, then remove the shade and/or bulb. While supporting the fixture so it does not drop, remove the mounting screws holding the base to the ceiling or wall, then carefully pull the fixture free.

In porcelain fixtures, you probably will find circuit cable or switch conductors fastened directly to terminal screws in the base (the hot conductor should be fastened to a brass or copper-colored screw near the center of the fixture, and the neutral conductor should be fastened to a silver-colored screw nearer the rim on the fixture's socket). In metal fixtures, the conductors should be spliced

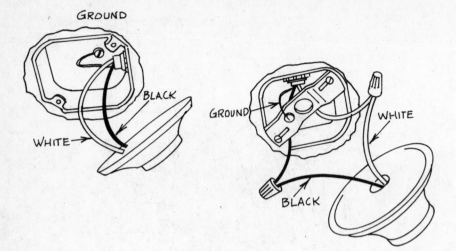

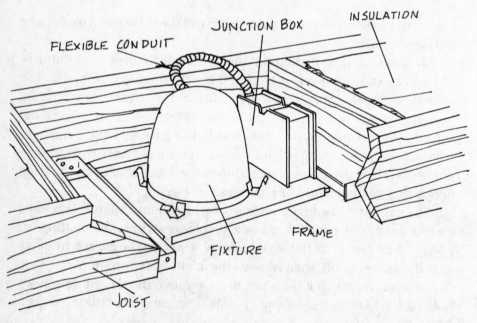

Illustration 95. Recessed ceiling light fixtures

with wire connectors to color-coded pigtails (black for hot and white for neutral). Before touching any wiring, use the voltage tester again to confirm that the power is off by touching the probes first to the conductors (if these are spliced to pigtails, insert the probes into the wire connectors to contact the conductors' bared ends) and then to each conductor and the outlet box.

Assuming that the power is off, examine the fixture for damage and signs of wear, and clear it of any debris. Renew the connections if they are burned or nicked, then reattach it—unless you plan to test it further as described below. Fixtures installed before 1960 might have been attached directly to the outlet box behind or above them. To comply with current NEC recommendations, you should upgrade such an installation by attaching a metal mounting strap (these are available at hardware stores) across the outlet box and securing the fixture to the strap instead of the box. This results in a stronger connection.

Central-mounted fixtures and chandeliers are attached to a threaded spindle found in some boxes. These do not require mounting straps. Outlet boxes for fixtures weighing over fifty pounds must be secured to a joist or cross brace. If you find no outlet box enclosing the fixture wiring (fairly common in ceilings of prewar houses), one must be installed.

Recessed incandescent fixtures often overheat, especially if ceiling insulation has been incorrectly installed too close to them. An airspace of at least three inches should surround ordinary recessed fixtures, and they should never be covered. To retain attic insulation, replace ordinary recessed fixtures with fixtures rated by Underwriters Laboratories for direct insulation contact. These are labeled "I.C." (insulated ceiling) inside the bulb housing.

Fluorescent and track lighting fixtures are mounted against outlet boxes in the same manner as incandescent fixtures. Older installations should be upgraded to include a mounting strap. Circuit cable conductors pass through the housings of some fluorescent fixtures, allowing the connections to be tested and examined after removing the tube (bulb) but without requiring that the entire fixture be freed from the ceiling. Some track lighting fixtures may be designed this way also, but to save space (actually, to preserve a trim appearance) the connections on most are located inside the outlet box, which means that the fixture must be removed to inspect them.

Other Considerations

Besides being in good condition, household wiring and equipment must meet certain standards of capacity to be considered safe. While local electrical codes are the final authority, the National Electrical Code recommends the following:

- Branch circuits supplying ceiling or other hard-wired lighting fixtures should be separate from circuits supplying receptacles. The normal amperage rating for a lighting circuit is fifteen amps. This is considered sufficient to supply three watts of lighting per square foot; by this calculation individual lighting circuits should serve no more than five hundred to six hundred square feet of floor space. Multiple circuits should be used to supply lighting on each floor and, if possible, each room. This way a failed circuit will not plunge large areas of a house into darkness. The conductor size of lighting circuits must be fourteen-gauge or larger (REMEMBER: gauge size decreases as conductor size increases).
- There must be at least one wall switch for operating lighting in each room, including the garage, attic, basement, and at outdoor entrances. Three-way switches should be installed at each end of corridors and at the top and bottom of stairs. Locate switches near doorways, on walls not covered when doors are open. Standard height for switches is about forty-eight inches.
- Branch circuits supplying outlets for general use may be rated for fifteen amps or twenty amps, but conductor size must be twelve-gauge. Fifteen-amp circuits may power floor and desk lamps, and other relatively light loads such as televisions, household vacuum cleaners, stereos, and small hair dryers; individual circuits should contain no more than nine receptacles. Twenty-amp circuits can handle heavier loads such as washing machines and portable electric heaters; individual circuits can contain up to twelve receptacles.
- Receptacles should be placed at eight-foot intervals along walls, twelve to eighteen inches above the floor to avoid kicking

plugs. No space along a wall should be more than six feet from a receptacle.

- Even small wall areas at least two feet wide should have receptacles, and there should be at least one receptacle for outdoor use. Outdoor receptacles should be GFCI models, housed in weatherproof outlet boxes with covers.
- Kitchens should contain at least two twenty-amp circuits for appliances. The circuits should be arranged so their receptacles alternate along walls; this way there is less chance of operating two appliances simultaneously on the same circuit. Receptacles should be installed along any counter twelve inches wide or wider.
- Critical-use stationary appliances, especially refrigerators, freezers, water pumps, furnaces, and heavy duty exhaust fans, must have individual, or *dedicated*, circuits supplying no other appliance. However, refrigerator and freezer circuits usually may also carry an electric clock to serve as a signal in case the circuit fails.
- GFCI receptacles only are allowed in bathrooms, on kitchen counters within six feet of a sink, in garages, outdoors, and in other locations where risk of electric shock is great. There must be at least one GFCI receptacle in the basement.

Mapping Circuits for Safety

Analyzing your home's electrical system is easier and making repairs is safer if you make a map of the electrical circuits for reference. With a circuit map you can know at a glance which circuits serve which outlets, and you can easily determine whether any circuits supply too many outlets and therefore are overloaded.

Creating a circuit map is easy, but is also tedious and fairly time-consuming: mapping an entire house can take several hours. Essentially, the process involves turning off one circuit at a time and checking to see which receptacles, light fixtures, and appliances it supplies.

Start by drawing a floor plan of the home on notebook-size sheets of paper. The plan may consist of individual floor diagrams or of dia-

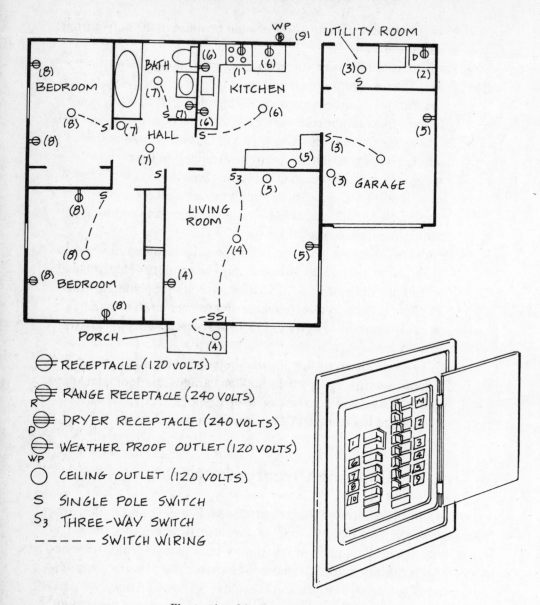

Illustration 96. Typical circuit map

grams of individual rooms. Include all areas served by electricity, including hallways, closets, attic, basement, garage, and outdoors. Drawing to scale is not necessary, but make the diagrams large enough so that you will have room to add notations and symbols. Write the name of each room or area inside its outline.

Next, indicate the location of all electric outlets—receptacles, fixtures, switches, permanent appliances, doorbells, thermostats, fans, everything. Don't overlook light fixtures in closets and other obscure areas. Distinguish between 120-volt outlets and 240-volt outlets. The latter, if they are receptacles, are clearly marked and have slots that will not accept ordinary appliance cord plugs. Appliances wired directly to a circuit should have a nameplate with voltage information printed on it. When showing switches, draw dotted lines linking them with the outlets they control. As you leave each room, turn on several items. When you are finished, go to the service panel and attach squares of durable adhesive tape to each fuse or circuit breaker controlling the branch circuits.

Begin mapping by tripping the breaker or removing the fuse of one circuit. Then find the room or area to which power has been shut off; that is, where items you switched on earlier are now off.

When you find the room, test all the outlets nearby. Operate switches to check light fixtures; to test receptacles, plug in a small desk lamp or use a voltage tester. Be sure to test both sockets in a receptacle in case they are served by separate circuits. Check other areas besides the room in question; usually circuits serve areas on several floors. If you cannot find outlets that no longer work, check individual appliances that may be served by dedicated circuits.

Assume that all nonfunctioning outlets you find are supplied by the disconnected circuit. On the map, write the number of the circuit next to the outlet locations you drew earlier. When you think you have recorded all nonworking outlets, restore the circuit and check your work; the outlets that you recorded should now function. If you find any that do not, circle them on the map.

Repeat the mapping process for all of the remaining circuits. When you are finished, all the outlets on the plan should have a number corresponding to a circuit. If any remain unidentified, use the process of elimination to determine the circuit supplying each one. Outlets that do not function when power is restored may have been disconnected during an earlier renovation, or there might be a problem that should be inspected by an electrician.

Make a photocopy of the map and keep it in a plastic folder attached to the service panel. Keep the original in a safe place.

Do You Need More Circuits?

You can use a circuit map to determine whether circuits are over-loaded. For each circuit, simply examine every load connected to it and write down the amount of its wattage, labeled "Watts" or "W." Light bulb wattages usually are printed on top of the bulb. On appliances and electric motors, wattages are either stamped onto the item or are printed on a label or nameplate. If instead of wattage you find an amperage rating (labeled "Amps" or "A"), multiply the amount by the voltage of the circuit, either 120 or 240. Add the wattages for all the items.

Calculate the maximum amount of wattage the circuit can carry by multiplying the circuit's voltage by the amperage rating of its fuse or breaker. For example, a 120-volt circuit protected by a 15-amp fuse or breaker has a maximum wattage of 1,800. According to the National Electrical Code, demand on a circuit must not exceed the circuit's "safe capacity," usually defined as 80 percent of its maximum wattage (this is found by multiplying the maximum wattage by .80). In the example above, 1,440 watts is the safe capacity of a circuit whose maximum is 1,800 watts. Therefore, if the total wattage of all the items on such a circuit is more than 1,440, the circuit should be considered overloaded. The solution is to move some items to a different circuit or to operate only a few items—their total wattage must be less than 1,440—at the same time.

Of course, if the house lacks enough circuits to allow spreading electrical loads evenly among them, more are needed. They can be added fairly easily provided the service panel contains room for additional fuses or breakers. Even if it doesn't, there may be solutions short of installing a larger service panel—such as adding a supplementary panel (called a subpanel); consult an electrician for more advice.

You can also determine the need for more circuits by estimating your home's total power needs based on accepted electrical industry standards:

1. Figure your home's square footage: if you are considering adding living space, include this amount also. Then multiply the

area by 3 watts. This is the average wattage supplied by 15-amp circuits in an ordinary house.

_____ sq. ft. x 3 = _____ watts

2. Next, add together all 20-amp appliance circuits (including any that are planned) and multiply the sum by 1,500 watts.

_____ 20-amp circuits x 1500 = _____ watts

3. Add wattages of appliances served by dedicated circuits (for example, washer, dryer, garbage disposer, refrigerator; do not include furnace or air conditioner).

_____ watts

4. Total the wattages above and then subtract 10,000 watts. Multiply the difference by .40. (The NEC uses this method to derive an average wattage representing the fact that not all outlets in a home are used at the same time.)

total wattage = _____ – 10,000 watts = _____
_____ x .40 = watts

5. Add the figure from the step above to the 10,000 watts subtracted earlier.

_____ watts + 10,000 watts = _____ watts

6. Add to the figure from the step above the wattage of the heating system or air conditioner, whichever is greater.

_____ watts + _____ watts =
Grand Total _____ watts

7. Divide the Grand Total watts by the voltage of the household service (240 volts) to find the total amperage required.

_____ Grand Total watts ÷ 240 = _____ amps

The answer in amps represents your home's total power needs. Compare this figure to the amperage presently available (to do this, add the amperages of all fuses and circuit breakers in the service panel). If the power needs of your home are greater than the amperage presently available, more circuits are needed. You can estimate how many by dividing the amount of the overage by either fifteen or twenty, depending on the amperage of circuits you would install to remedy the condition.

Avoiding Falls
and Other Injuries

HOMES CAN AND OFTEN DO CONTAIN DANGEROUS CONDITIONS other than fire hazards that heighten chances for accidents such as falls, electric shock, burns, and poisoning. Even a little remodeling usually can make a home safer, and the results can mean the difference between happy, convenient living and tragedy. Often, simply rearranging furniture is enough. What follows here are general safety hints with universal benefit. (Children and older people have special needs that are covered fully in the chapters that follow—"Childproofing" and "Creating a Barrier-Free Home.")

As was true for locating fire hazards, the best way to discover household safety hazards is to tour your home in search of them. As you go from room to room, make an inventory of places where accidents are most likely to occur—stairways lacking railings, bathrooms with unprotected electrical outlets or slippery tubs, cramped kitchens, ill-lit basements, and the like. Ask family members or other household occupants to identify what each believes to be the least convenient or most hazardous feature in the house and to recall past accidents and near misses. Write down every item, prioritize them all, and then set about to remedy each condition, starting with the most dangerous.

Electrical Hazards

Electrical hazards usually are high on the priority list, especially in older houses. These hazards are described in the chapters "Pro-

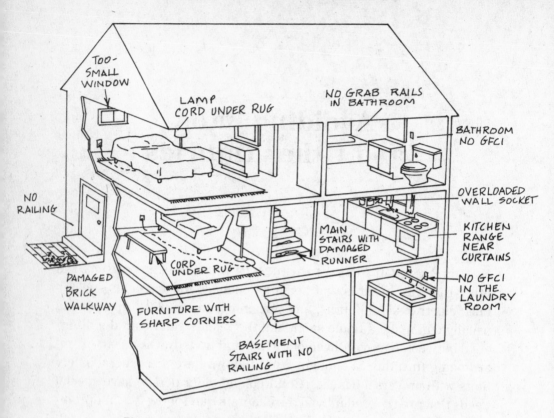

Illustration 97. Household accidents waiting to happen

tecting Your Home from Fire" and "Upgrading Old Wiring." Replacing receptacles (wall outlets), even if they are sound, in potentially damp areas such as kitchens, bathrooms, and laundry rooms with ground-fault circuit interrupters (GFCIs; see "Upgrading Old Wiring") is one of the most important steps you can take to reduce the risk of electric shock and even electrocution. To further reduce risks, replace any metal wall plates surrounding receptacles with plastic ones, which cannot become electrically charged. Naturally, replace any visibly hazardous electrical fixture or appliance, or have it professionally repaired, and keep electrical cords from beneath furniture and rugs.

Falls

Falls can cause serious injuries and may account for at least 40 percent of household accidents, according to a 1982 study undertaken by the Buffalo Organization for Social and Technological Innovation and sponsored by the federal government. Keep traffic lanes through rooms free of obstacles, especially low furniture. Wax floors only when necessary to protect their surfaces. Make sure rugs do not slide underfoot. If they do, secure them with a carpet liner underneath (liners also reduce wear on rugs and carpets) or fasten them to the floor with double-sided carpet tape. Choose rugs, carpet, and stair runners in solid colors and made of short, dense pile. Avoid splashy or busy designs and long shag pile; patterned rugs can be confusing, and long pile is slippery. Remove any rug or carpet that is worn through.

Staircases

Stairs deserve careful scrutiny; the study cited above determined them responsible for more than half of the falls reported. Most building codes regulate staircase designs and features; typical minimum dimensions are a width of thirty-six inches for main staircases and thirty inches for secondary stairs; a riser height (the vertical distance between steps) of six to seven inches; a tread depth of nine to eleven inches; and headroom of six feet eight inches (some codes allow basement stairs to have only six feet six inches of headroom). Sturdy handrails are required; they usually must be between thirty-four and thirty-eight inches above treads, measured in line with the risers.

Unfortunately, little can be done with stairs that do not meet code standards unless you are prepared to replace them, an expensive proposition and not always possible. Nevertheless, you can make stairways safer by checking to see that all parts are sturdy and securely fastened, by making sure lighting is adequate but not glaring (to avoid having to negotiate stairs in the dark, light switches should be installed at both the top and bottom of stairwells, at least thirty-six inches away from the initial step), and by installing nonslip footings and handrails that are easily gripped for support.

Repair weak or broken stairs and loose banisters, or hire an ex-

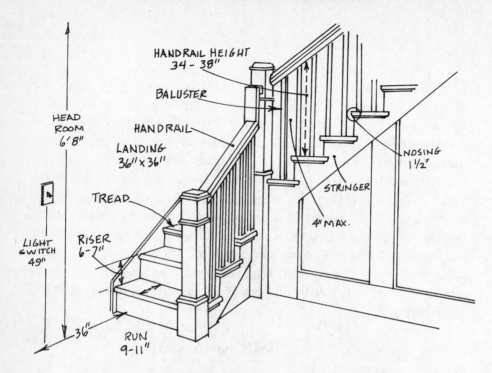

Illustration 98. Code-approved stairs and railing

perienced finish carpenter to perform the task. Staircases are complicated—easy to mar and often difficult to repair—because their joints usually are deliberately concealed. Squeaking stairs sometimes signal problems that may worsen over time; pay prompt attention to cracks, especially along the nosing (the overhanging lip of treads). A broken nosing can cause a fall and may necessitate replacing the entire tread.

To repair a cracked nosing, pry the sections apart slightly with a putty knife and then work glue between them with a playing card or some other thin applicator. Wiggle the loosened piece gently as you apply the glue, and then press the pieces together until an even line of excess glue oozes along the length of the seam. Then drill holes and drive long finishing nails at angles into the edge of the nosing to hold the split together. Wipe away the excess glue with a damp cloth.

Paint the top and bottom stairs in a basement so they are easy to see, especially if a door at the top of the stairs opens directly onto the staircase and not onto an elongated step or landing. On stairs with

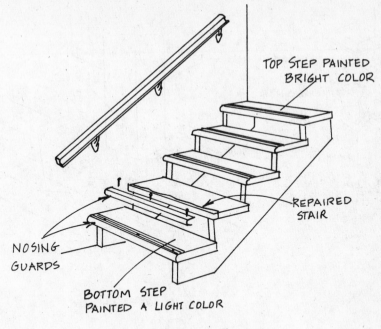

Illustration 99. Stair repairs

bare treads, install rubber or metal nosing guards that enhance traction when you are climbing and descending. These are easy to install using nails, screws, or contact adhesive. When installing fasteners, drill pilot holes for them first to prevent the wood from splitting.

Another solution to the problem of slippery stairs is to install a stair runner. Use short, dense carpeting as was described earlier. Shift the runner before the areas covering the tread nosings become worn.

Banisters

Loose banisters, or staircase railings, can be the result of loose newel posts, the large posts at the head and foot of a staircase. Solid newel posts in newer houses usually are bolted to the stair carriage (the notched, diagonal board supporting the treads) and to a joist underneath the floor. To locate the first bolt, look for a wooden plug or a removable cap on the outside of the newel; if the plug is glued into the newel you will have to chisel it out to reach the bolt.

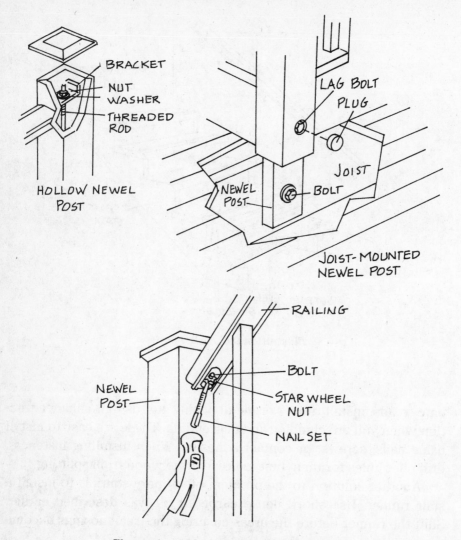

Illustration 100. Newel posts and railing

To find the second bolt, search overhead from the basement. You should find the base of the newel post extending vertically through the subfloor and across a joist. Use a large socket wrench to tighten both fasteners. Replace any that cannot be tightened.

Similar bolts may also secure an upper newel post. Unfortunately, reaching the lower bolt may require removing a portion of the ceiling beneath it.

In older homes, newel posts are often hollow and rest on top of

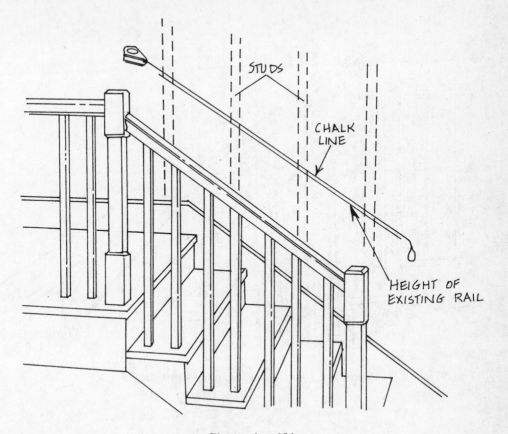

Illustration 101.
Installing a handrail: Step 1

the subfloor instead of passing through it. Inside the post is a horizontal bracket through which is threaded a long, vertical steel rod
that bolts the post firmly to the floor.

To access the bracket, remove the top of the post by tapping it
with a mallet. Tighten the nut on the end of the rod with a wrench.
Also inspect the bracket and tighten any fasteners that secure it to
the post.

If the newel posts are tight but the banister still wobbles, examine
the handrail. Most are built of segments joined end-to-end by steel
pins threaded at each end. Look underneath the rail for holes that
allow tightening of the pins with a hammer and either a nail set or a
screwdriver to turn the pins' star-shaped nuts. (The holes might be
plugged, as was described earlier, and require chiseling to open.) To

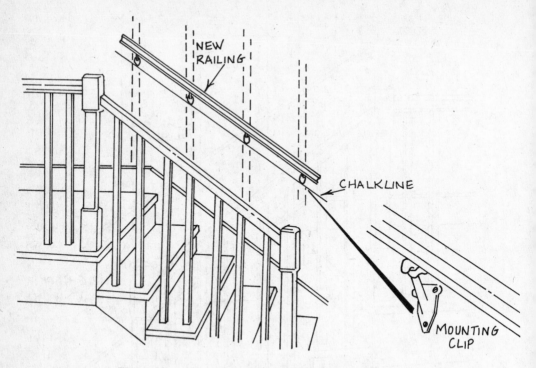

NEW RAILING

CHALKLINE

MOUNTING CLIP

Illustration 101A. Installing a handrail: Step 2

tighten a nut, place the tip of the nail set or screwdriver against one of the points of the nut, then strike the other end of the tool with the hammer to turn the nut clockwise. Loose handrail spindles can also weaken banisters and cause them to wobble. Unless these can be reglued easily, repairing them generally is best left to a professional. Don't be tempted to secure loose banister parts with nails or screws; these eventually will loosen.

Stairs having only a banister on the outside can be made safer by attaching an auxiliary handrail to the wall. Buy a wooden handrail at a building supply store or lumberyard; choose a smooth, round rail one and a quarter or one and a half inches in diameter (no larger) that can be gripped firmly. Also buy handrail mounting brackets and clips. These fasten to the underside of the rail; be sure they will hold the rail at least one and a half inches away from the wall. Buy enough brackets and clips to install at thirty-two-inch intervals and at each end of the rail.

To position an auxiliary handrail, measure vertically between the surface of a tread and the bottom of the existing handrail, then subtract the height of a bracket. Mark this height on the wall at the top, middle, and bottom of the staircase, and then lightly snap a chalk line between the points.

Locate the studs in the marked wall. Use an electronic stud-finder so you can mark their thickness by drawing vertical lines crossing the chalked line. Align mounting brackets by placing a bracket's bottom end on the chalked line and moving the bracket along the line so that at least two of the mounting holes are over a stud. Mark the locations of the holes on the wall by drilling pilot holes for screws using the holes in the brackets as guides. If some holes do not lie over a stud, insert hollow-wall anchors for screws in these holes. Then fasten the brackets securely to the wall with sturdy wood screws (screws entering studs should be at least two inches long).

To mount the handrail, rest it on the brackets (have a helper at the foot of the stairs hold it in place there). Fit a mounting clip around a bracket and against the underside of the rail; then use an awl or a sharp pencil to mark through the holes in the clip onto the rail's surface. Repeat the process at each bracket, being careful not to let the handrail slide or rotate. When you are finished, take the handrail down, drill pilot holes for screws at each mark, then attach all of the mounting clips loosely with the screws that come with them. Reposition the handrail by sliding all of the clips over the brackets; finish securing the rail by tightening the screws fully.

Bathroom Safety

Bathrooms are the sites of many household falls and other accidents. The U.S. Consumer Products Safety Commission estimates that 200,000 bathroom-related injuries occur each year in the United States, with falls accounting for the majority (perhaps 75 percent of the total), and hot water burns next in line.

Grab bars (similar in purpose to handrails), therefore, are valuable additions to bathroom decor. You do not have to settle for stainless steel models, either, which some homeowners dislike because of their institutional appearance. Increasingly, grab bars are available with colored enamel finishes that match or complement almost any color scheme.

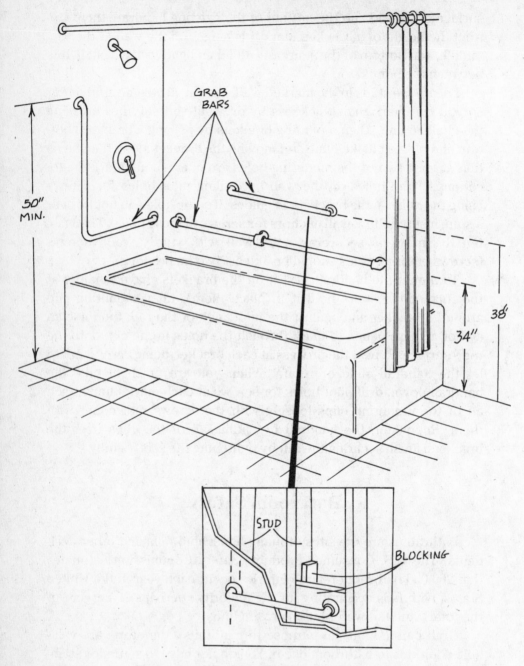

GRAB BARS

50" MIN.

38'

34"

STUD

BLOCKING

Illustration 102. Installing bathroom grab bars

To prevent falls in a tub or shower, mount grab bars on the walls above the tub. There are several recommended configurations. For climbing into a tub or shower, a vertical bar affording a grip forty to fifty inches above the floor on one or both of the tub's end walls is best. For a bathtub with a shower, also install two horizontal bars thirty-four and forty-eight inches above the tub floor. For a tub only, install the two additional bars vertically, equidistant from each other and the tub's end walls at a height that offers a grip between thirty-four and forty inches above the tub floor. For a shower only, install only one supplementary bar vertically on the inside wall of the enclosure at the same height as the bar or bars outside.

It is important that grab bars be fastened to studs, not with hollow-wall fasteners, which lack adequate holding strength. You can locate wall studs behind an ordinary wallboard wall easily with an electronic stud-finder, but if the wall is covered with ceramic tile you might have to drill test holes instead. Choose a place where the holes can be easily filled; near the top of a wall if possible, otherwise along grout joints.

You may have to drill through ceramic tiles to mount grab bars. To do this, cover the tile with masking tape and mark where you want to drill. Drill through the tile with an electric drill equipped with a ceramic-tile bit and then into the stud using an ordinary twist bit. Remove the tape. Mount the grab bars with long, stainless steel screws (these won't rust); tighten them securely but be careful not to overtighten, otherwise you may crack the tiles underneath. If installing wall-mounted grab bars simply is not feasible, use a grab bar that fits over the side of the bathtub; it is better than nothing at all. Never rely on soap dishes or towel racks as grab bars; they seldom are adequately fastened. As an additional precaution, install nonslip strips or a rubber mat on the bathtub or shower floor.

To prevent scalds, the easiest step to take is turning down the thermostat on the household water heater to 120 degrees Fahrenheit. (Many homes and apartments have hot water settings as high as 160 degrees; even at 140 degrees, only three seconds of exposure are needed to produce third-degree burns on sensitive skin.) To test the temperature of the water, place an oven thermometer in a cooking pan with a handle and run hot water into it until the thermometer reading stabilizes.

Also recommended are pressure-balancing tub and shower con-

Illustration 103. Pressure-balancing faucet control

trols. Relatively new, these replacement single-handle faucets prevent bursts of hot or cold water caused by using water elsewhere in the house—for example, when a toilet is flushed or a washing machine is turned on. Most building codes now require pressure-balancing controls in new construction, including remodeling.

Although pressure-balancing controls look like any other single-handle shower or tub faucet, they contain a special diaphragm or piston inside that moves with changes in water pressure to instantly

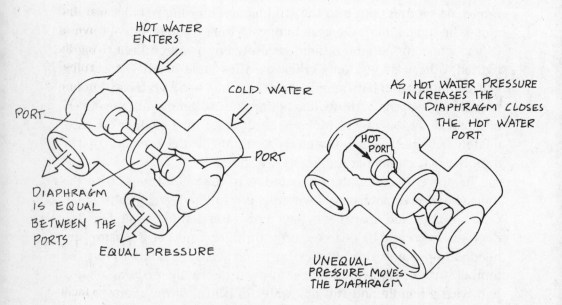

Illustration 103A. How a pressure-balancing faucet works

balance the pressure of the hot and cold supply lines. Most reduce hot water flow to a trickle if the cold water supply suddenly fails; during use, water temperature is maintained within two or three degrees. As further protection, most pressure-balancing controls have an adjustable stop that limits the amount of hot water the valve can deliver.

Installing a pressure-balancing control usually is a job for a licensed plumber. Soldering—which requires practice and special skills—is necessary. Also, modifying bathtub and shower plumbing typically is fraught with minor problems that can be solved easily with a professional's tool kit and plenty of spare parts but can be unconscionably tiresome and time-consuming without these aids. Prices of pressure controls span as wide a range as prices of most plumbing fixtures, from around $100 to upward of $500. For a no-frills model that will last many years, expect to pay around $150.

In 1992 the National Kitchen and Bath Association (NKBA), a trade organization, announced safety and design recommendations for residential bathrooms, the first such guidelines since the ones developed by the U.S. Department of Housing and Urban Development during the 1950s (which still form the basis of many local building code regulations concerning bathrooms). If you are considering remodeling a bathroom or adding a new one, you should obtain and study them; besides recommending that tubs and showers have at least one grab bar to facilitate entry and that pressure-balancing devices be installed to prevent scalds, some twenty-five other "rules" are presented. Following is the complete list of twenty-seven recommendations:

1. Entryways should be at least thirty-two inches wide.

2. No doors should interfere with fixtures.

3. Provide adequate ventilation.

4. All electrical receptacles must have ground-fault circuit interrupters. Do not install switches within sixty inches of any water source. Use a moistureproof, special-purpose light fixture above tub or shower units.

5. Provide at least six inches of floor space between fixtures to facilitate easy cleaning.

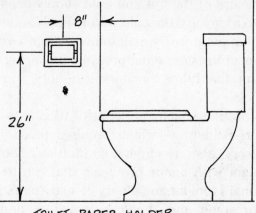

TOILET PAPER HOLDER

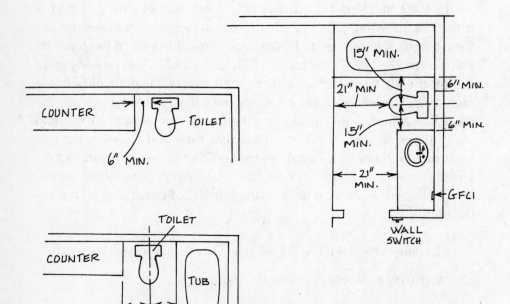

Illustration 104. Toilet clearances

6. Provide at least twenty-one inches of open space in front of lavatories.

7. Provide at least twelve inches of clearance from the centerline of a lavatory to a side wall.

8. The minimum clearance between side-by-side lavatories should be thirty inches measured from the bowls' centerlines.

9. Provide at least fifteen inches of clearance from the center of a toilet to any obstruction or fixture on either side.

10. Provide at least twenty-one inches of open space in front of a toilet.

11. Install a toilet paper holder within easy reach of the person seated on the toilet—ideally, slightly in front of the edge of the bowl and twenty-six inches above the floor.

12. Provide at least fifteen inches of space from the center of a bidet to any obstruction, fixture, or equipment on either side.

13. Provide at least twenty-one inches of open space in front of a bidet.

14. Provide storage for soap and towels within reach of a person seated on a bidet.

15. No more than one step should lead to a tub. That step must be at least ten inches deep and must not exceed seven and a quarter inches in height.

16. Bathtub faucets should be accessible from outside the tub.

17. In baths with whirlpool tubs, provide access to the motor.

18. Install at least one grab bar to facilitate bathtub or shower entry.

19. Shower enclosures should measure at least thirty-two inches by thirty-two inches.

20. Shower enclosures should include a bench or footrest.

21. Provide at least twenty-one inches of open space in front of a tub or shower.

22. A shower door should swing into the bathroom, not into the shower. This allows easy access to a person who is injured while inside the shower enclosure.

23. Equip showerheads with a pressure-balancing temperature regulator or temperature-limiting device to prevent scalds.

24. Install slip-resistant flooring.

25. Provide adequate storage including a counter or shelf around the lavatory, space for grooming equipment, a shampoo and soap shelf in the shower and/or tub, and hanging space for linens.

26. Provide adequate heat.

27. Provide adequate general and task lighting.

For additional comfort and safety in showers, showerheads should be between seventy-two and seventy-eight inches high, and shower doors should be made of shatter-resistant safety glass (older doors may not be shatterproof).

Kitchen Safety

The NKBA has drawn up similar guidelines for kitchens. Many focus on convenience—minimum countertop and cabinet space, for example—but others specifically address safety. Those pinpointing safety issues are:

- Entryways should be at least thirty-two inches wide.
- Doors should not interfere with work centers, appliances, or counters.
- All receptacles within six feet of a water source should be protected by ground-fault circuit interrupters.
- A cooking surface should not be located below an operable window unless the window is at least three inches behind the appliance and twenty-four inches above it.
- Space next to or above an oven should provide at least fifteen inches of clearance if the appliance door opens into a kitchen traffic area.
- Space next to, above, or below a microwave oven should provide at least fifteen inches of clearance for opening the door.
- All surface cooking appliances should be adequately ventilated.
- Clearance between a cooking surface and a protected surface

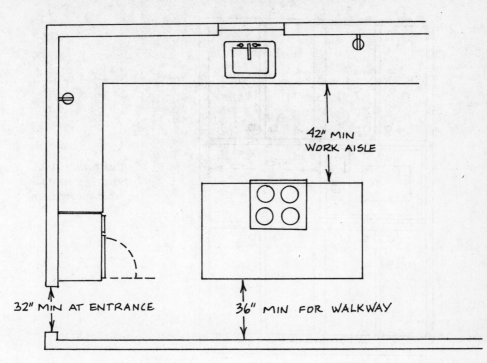

42" MIN
WORK AISLE

32" MIN AT ENTRANCE

36" MIN FOR WALKWAY

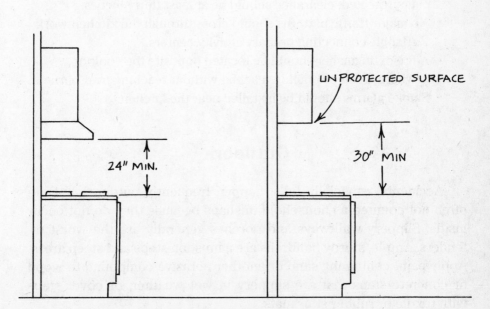

UNPROTECTED SURFACE

24" MIN.

30" MIN

Illustration 105. NKBA kitchen safety recommendations

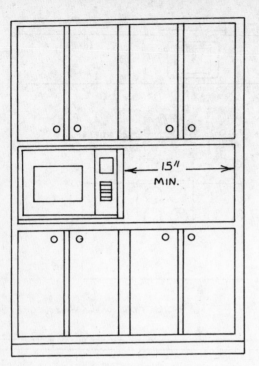

*Illustration 105A.
Safe clearance for
microwave oven*

above should be at least twenty-four inches; if the surface is
unprotected, clearance should be at least thirty inches.

- No major traffic patterns should cross through the kitchen work
triangle connecting primary activity centers.
- A fire extinguisher should be located opposite the cooktop,
where it can be quickly reached without reaching over burners.
- Smoke alarms should be installed near the kitchen.

Outdoors

Accidents, especially falls, happen frequently outdoors but are
often not counted as household mishaps because they do not occur
inside. Slippery walkways and porches generally are the worst offenders. Ample, sturdy handrails are a must on steps and steep areas.
Apply paint containing sand or another abrasive compound to wood
or concrete steps that are slippery in wet weather, or cover steps
with textile or rubber stair mats.

Installing outdoor lighting increases security as well as safety.

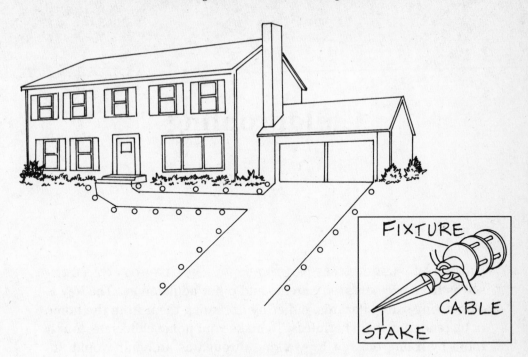

Illustration 106. Low-voltage outdoor lighting

Low-voltage lighting is inexpensive and hardly more difficult to install than lighting for a Christmas tree. Kits come with instructions; essentially, the process consists of mounting a transformer (supplied with the kit) where it can be plugged into an outdoor receptacle, and then laying out a cable along the path chosen for the lighting fixtures. These fixtures are attached to the cable, usually by squeezing special clamps that pierce the sheathing to make contact with the wires inside, and then are driven into the ground (if they are on stakes) or are fastened with brackets to objects such as deck railings or porch steps. The cable then can be tucked out of sight beneath shrubbery or buried a few inches deep to conceal it.

Low-voltage lighting is so safe that some manufacturers recommend assembling layouts with the power on to permit detecting faulty connections as you go. However, do not handle live low-voltage wiring if you wear a pacemaker; even the small amount of electricity the wiring carries can upset the device.

Childproofing

CHILDREN—ESPECIALLY TODDLERS—CANNOT PROTECT THEM-selves from accidents, so parents and other adults must. The key is eliminating safety hazards, either by excluding them from the home or by rendering them harmless. To make your home child-safe, tour it room by room, but not by walking through as an adult would. Instead, crawl from place to place on your hands and knees, or at least stoop to the floor in every room. This way you will see the world from a child's viewpoint, and by doing so you will detect many more dangers than you would by other means.

As is true for most safety tours, the kitchen is a good place to begin. Much of a child's time may be spent in that room and there are many potential hazards. First on the list are dangerous substances such as detergents, household cleaners, and other toxic items that are often stored in cabinets near the floor. Anything remotely harmful should be shelved, above a child's reach, in cabinets that can be locked.

Even so, childproof latches on every undercounter cabinet door and drawer are a must. These are inexpensive and easy to install. Most consist of a tough plastic tab that attaches to the inside of the door, usually near the top. The end of the tab fits around a screw installed so that it projects from the inside of the cabinet. The length of the tab permits the door to be opened an inch or so; not enough for a child to reach in but just enough for an adult to insert a finger and depress the tab to open the door fully. Although some children may eventually discover the latch, flexing the stiff plastic from which most latches are made requires adult strength.

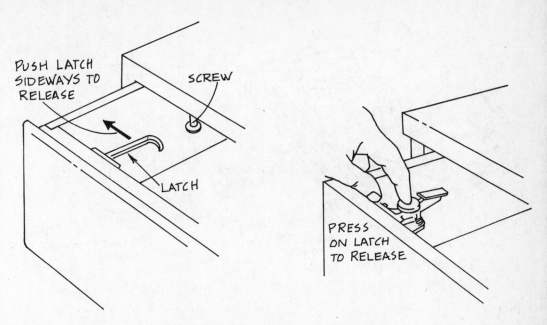

Illustration 107. Childproof latch and lock

Another means of preventing small children from opening double cabinet doors whose knobs are side by side is to slip a strong rubber band over them. Slip-on cabinet "locks" based on this idea are also available in stores.

Keep kitchen counters, backsplashes, and walls free of knives and other sharp objects, and of appliances with sharp cutting blades. Install switchplate locks on wall switches controlling a garbage disposer or other appliance (switchplate locks require you to push a tab while moving the switch, making them too difficult for a child to operate). If the edges or ends of counters extend beyond base cabinets (so that a child's head might bump into them), attach foam rubber strips to the corners or lower edges of the counters using double-sided tape. Ready-made corner protectors are also available at many home centers.

Evaluate safety conditions above counter height also. Older children—and even resourceful toddlers—will climb to get at anything that attracts them. Install childproof latches on wall cabinets containing glassware, drugs (including vitamins), and any other items that should be off-limits. Keep knickknacks to a minimum, and

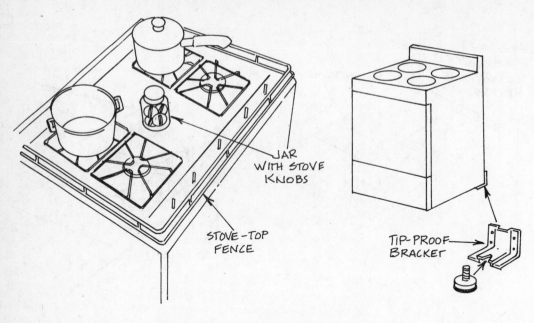

JAR
WITH STOVE
KNOBS

STOVE-TOP
FENCE

TIP-PROOF
BRACKET

Illustration 108. Childproofing a stove

never place anything that might attract a child on shelves above a
stove.

Speaking of stoves, these should be anchored to the floor with security brackets—metal fittings attached to the floor with screws. A
slot in the bracket accepts the foot of a range, preventing the appliance from tipping over. Security brackets can also be used to anchor
refrigerators. In addition, equip stoves with burner fences, protective
bars that encircle the stovetop, preventing small hands from reaching
pots and burners. Some parents also remove control knobs from
stoves and keep them in a nearby jar when the stove is not in use.

Finally, keep the floor clean and the garbage in a container beneath the sink (with childproof latches on the cabinet doors, of
course). Avoid leaving pet food, pet dishes, or litter boxes unattended
on the floor.

Check out bathrooms next. Like kitchens, they too hold fascinat-
ing items for a toddler or young child, and the contents of bathroom
cabinets can be very dangerous.

Simply throw away old drugs and medicines you no longer use.
Keep those you do, as well as other items such as scissors, razors,

and makeup (which can be toxic if eaten), in a locking cabinet. As was mentioned in the previous chapter, take steps to avoid scalds by lowering the temperature of the hot water in your home to 120 degrees and by installing pressure-balancing tub and shower controls. Install privacy locks on bathroom doors that can be opened from the outside with a special key. Keep the key in a safe place where it can always be found if a child locks herself or himself in.

When touring living areas, be on the lookout for sharp-edged furniture such as coffee tables. Install edge protectors wherever necessary, and think about removing fragile or potentially dangerous furniture such as pieces containing glass or sharp metal. Anchor tall freestanding shelves to the wall with angle brackets to prevent them from tipping over if a child attempts to climb them.

Replace faceplates surrounding wall outlets with childproof covers that slide over the slots of sockets not in use. These are better than most plug-in outlet protectors, which can be swallowed if they are pried loose. Tape electric cords to the wall or floor with sturdy tape, such as duct tape, so children cannot play with them or become entangled in them. If windows are covered by venetian blinds, cut the loop in the end of the leveler cord and attach plastic caps to the separate ends. Attach the loops of vertical blind cords to tie-downs mounted on the floor or wall.

Naturally, keep anything that is potentially dangerous and that might attract a child well above the height that he or she can reach. In living rooms and dens, this means stereos, computers, and other electrical equipment; art objects on shelves or in cabinets; and matches and cigarette lighters on coffee tables. Minimize floor and desk lighting; substitute ceiling lights and fixtures mounted high on walls. If you own a floor-model television, keep the screen covered and locked when not in use; surround fireplaces and woodstoves with sturdy railings. Many houseplants, including common varieties, are poisonous if eaten. Your local poison-control center can supply you with a list. Do not grow or keep poisonous plants in your home.

In other rooms, be alert for bureau drawers containing unsafe items such as clothes packed in mothballs, which are poisonous. Childproof latches can be installed on these drawers, or mothproofed items can be relocated to a locking trunk or chest. Be sure that children's furnishings such as lidded toy chests are safe. Instead of

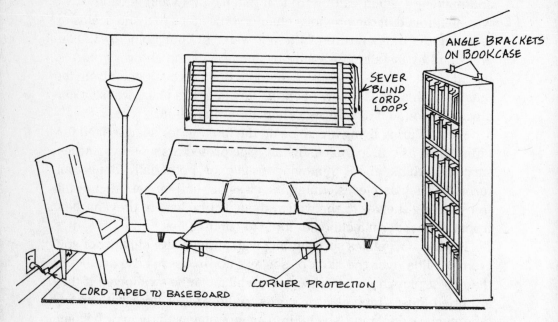

Illustration 109. Childproof living room

chests with hinged lids that may fall closed, use only chests with separate lids or sliding doors.

Keep clutter off tabletops and vanities, and teach older children to keep their rooms picked up for the safety of their younger siblings. To prevent children from wandering into closets, opening attic and cellar doors, or entering unoccupied rooms of others, use plastic doorknob protectors. These are clear plastic cylinders shaped like doorknobs that fit over the actual doorknob. Squeezing the protector while turning the knob permits opening the door. Children lack the strength required for this and also cannot perform both operations simultaneously. To keep a door locked, use a safety chain, barrel bolt, or simple hook and eye; install all of these devices out of reach of small children, but for safety do not install them too high for older children and adults.

Protect toddlers from stairways by installing gates at the top and bottom. These should consist of sliding panels with openings no wider than two and three-eighths inches; old-fashioned accordion-style gates can entrap a child's head. Choose a gate that is easy for

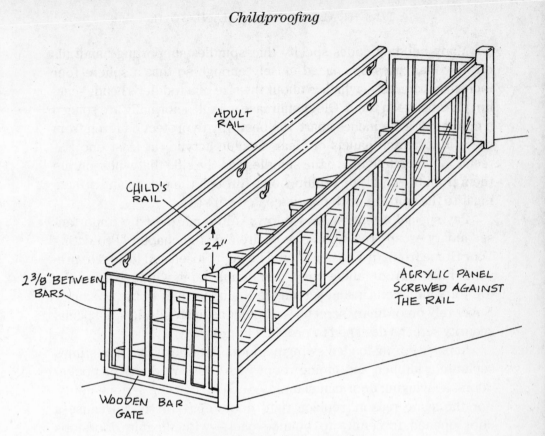

Illustration 110. Stairway safety

adults to open and close so it will be used consistently. Also, select a gate that installs with screws to a wall; pressure-mounted gates can be pushed loose.

Safety gates are certified by the Juvenile Products Manufacturers Association, a trade organization that regulates childrens' furnishings by issuing a seal stating that a product has met voluntary standards under the guidelines of the American Society for Testing and Materials.

Staircases must have handrails thirty-four to thirty-eight inches above the treads for adults. In a house with children, it also is a good idea to install supplementary handrails about twenty-four inches above treads; these should be designed and installed similarly, but should be of narrow stock that a child can easily grasp. For instuctions on installing handrails, see the chapter "Avoiding Falls and Other Injuries."

Many building codes specify that spindles supporting handrails on new staircases be spaced closely enough so that a sphere four inches in diameter—which is about the size of a toddler's head—cannot pass between them. Older staircase spindles normally are spaced a minimum of six inches apart; the best way to protect children from these is to attach panels of clear, flexible acrylic (at least one sixteenth of an inch thick) to the spindles (on the side that will separate them from the child) by drilling holes through the plastic and attaching it to the spindles with plastic-coated wire ties.

Pay special attention to windows. Although they are important secondary escape routes in case of fire (see the chapter "Protecting Your Home from Fire"), they should be kept closed and locked to prevent children from falling out. Open double-hung windows from the top for ventilation (install screens in the upper half of the opening). Never rely on ordinary screens to protect against falls; install special security screens designed to protect children.

Replace spring-loaded swinging doors like those that sometimes separate a kitchen and dining room with ordinary doors or pocket doors; a swinging door can slam into a child, causing serious injury. For the same reason, replace rigid doorstops—these can cause a door opened forcefully to bounce back—with flexible doorstops mounted near the tops of doors instead of near the bottoms.

Regularly inspecting and maintaining outdoor playground equipment is important—and often neglected. Keeping play equipment safe is paramount, but equipment that is well cared for will also last longer and be more enjoyable to use. If you are planning to install playground equipment or would like to compare your existing equipment with established safety guidelines, obtain a copy of Document F-1148-88, "Standard Consumer Safety Performance Specification for Home Playground Equipment," prepared by the American Society of Testing and Materials (ASTM). This may be ordered by contacting ASTM Customer Service, 1916 Race Street, Philadelphia, PA 19103. Cost at the time of this writing is $8. Your local building inspector may also have a copy (or a similar document).

Many parents and public health and safety experts are concerned about the possible hazards of the preservatives that are applied to wooden playground equipment. Unfortunately, conclusive evidence that the most popular method of preserving wood—pressure treatment with chromated copper arsenate (CCA)—is either safe or harmful when used in playground equipment is not yet available. Testing to determine this is currently under way by the Consumer Product Safety Commission. In the meantime, experts recommend that pressure-treated wood in playground equipment be sealed with water-repellent wood sealer for safety. Following is a checklist for maintaining playground equipment:

CARING FOR HOME PLAYGROUND EQUIPMENT

ITEM: Wooden playground equipment
CARE: Treat with water-repellent wood sealer at intervals recommended by manufacturer; smooth sharp edges with sandpaper, file, or electric router; remove splinters. Recess protruding ends of bolts by counterboring with a drill; alternatively, replace conventional nuts with capped, or "acorn," nuts.

ITEM: Metal playground equipment
CARE: Tighten loose hardware twice monthly during usage season; smooth away sharp edges with emery cloth or file; replace conventional nuts with capped, or "acorn," nuts; cover ends of hollow tubing with plastic or metal caps or inserts; lubricate moving parts with medium-weight general purpose oil or grease. Treat rusty areas with rust remover or rust converter (painted metal only).

ITEM: Swings
CARE: Test ropes or chains monthly with adult weight during usage season; smooth sharp edges of chain links; close any open hooks; make sure anchors remain secure. On swings attached to tree limbs, inspect rope closely and often for wear. When attaching new rope, thread it through sections of garden hose for protection where it passes over tree limb.

ITEM: Sandbox
CARE: Frequently rake and remove debris; dampen sand to prevent blowing in dry weather; cover sandbox when not in use to keep animals out; replace old sand with washed beach or river sand from building supply company.

ITEM: Ground cover

CARE: Keep playground area covered with a three- to six-inch-thick layer of shredded wood chips or bark (bark nuggets are not as resilient). Under swings, slides, and climbing equipment, increase depth of cover to nine inches or consider outdoor rubber matting.

Creating a Barrier-Free Home

COMPETELY REMODELING A HOME FOR THE CONVENIENCE OF elderly or physically disabled occupants takes thorough planning—and usually the skills of an architect and a contractor who specialize in such projects. For example, a common mistake is to spare no expense in remodeling a bedroom or building an addition to accommodate a disabled person, but then to neglect making other modifications—many of them simpler and less expensive—that enhance the person's access to the rest of the house, where most daily living and family activity take place. Such well-intentioned but misguided efforts can isolate the very person for whom the special accommodations are intended and actually make life for the person and caregiving by other household members more difficult.

This chapter describes a broad range of basic adaptations aimed at increasing overall accessibility in a home by eliminating common structural and design barriers. Like everyone else, disabled residents need to be able to enter and leave a building, move about inside, carry on activities, use the kitchen and the bathroom, and retire to a bedroom. Consult a local building inspector before taking on remodeling projects such as building a ramp or enlarging doorways; many must conform to special sections of local building codes. These sections often are based on the Uniform Federal Accessibilities Standards, the government's guidelines for public buildings.

Access

Practically speaking, access to a house begins at the edge of a driveway or another point of exit from a car. Ideally, the path from car to house is level, smooth, and at least three feet wide (for a wheelchair, four or five feet wide is preferable, because guiding a wheelchair in a straight line is difficult for a user turning the wheels by hand).

Generally, some incline, step, or stairway exists between a driveway or carport and an entry door, necessitating a ramp. The width of the ramp should be the same as that of the walk. Although building codes typically specify the slope to be one in twelve—that is, a one-inch rise for every twelve inches in horizontal length—a shallower ramp (up to one in twenty) is easier to manage whether walking or in a wheelchair and is required by some codes where snow and ice are common.

A poured concrete ramp is the most durable and the easiest to maintain, but a wooden ramp is usually less expensive and easier for an amateur to build, especially if the rise exceeds a foot. Construction can be similar to that of an outdoor deck, consisting of cedar, redwood, or pressure-treated lumber (all of which resist decay) rest-

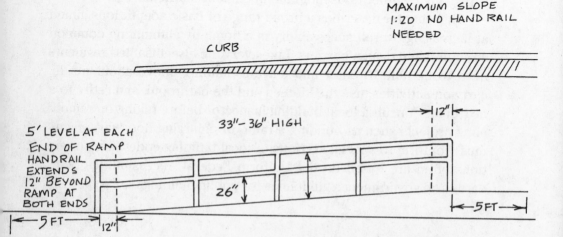

Illustration 111. Ramp design dimensions

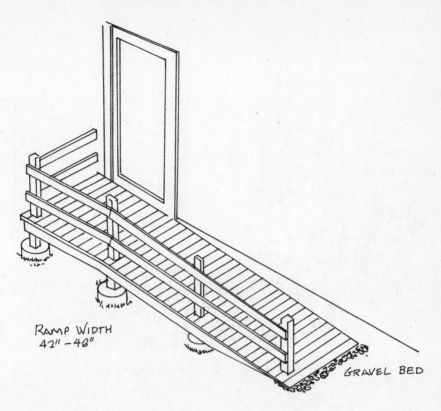

RAMP WIDTH
42" – 46"

GRAVEL BED

Illustration 111A. Wooden ramp

ing on foundation posts set into the ground. Generally, there must be a level landing where a ramp meets a doorway to allow turning the wheelchair when opening the door. A landing for a door that opens inward should be at least thirty-six inches wide; one for a door that opens out should be at least sixty inches wide.

A well-designed ramp can be inconspicuous. Where a ramp must be quite long to reach a high doorway, consider locating it at a side or rear entrance. Long ramps can consist of short segments built parallel or at ninety-degree angles to each other, like flights of stairs, and linked by landings.

All ramps must have a handrail unless their slope is shallower than one in twenty. For pedestrians, standard handrail height is thirty-six inches. For wheelchair users, a second rail should be added below it, at twenty-six inches. Handrails must be easy to grip; they should be no more than one and a half inches in diameter and should

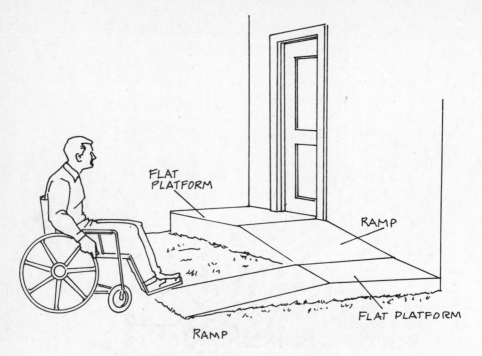

Illustration 111B. Space-saving ramp

extend at least twelve inches beyond the foot of the ramp. (For more on handrails, see the chapter "Avoiding Falls and Other Injuries.")

Doorways, whether for an entrance or for interior passages, should provide at least thirty-two inches of clear space to allow a wheelchair to pass. Some doors can be mounted on offset hinges so they swing out and away when opened, allowing full use of the doorway space. But if the door frame itself lacks the necessary width, major remodeling can be necessary to install a new door, because rough framing in the wall also may require replacement to accommodate a wider opening. Select the "handedness" of doors to provide at least eighteen inches of space adjacent to the latch on the door's inward-swinging side. This insures that the person opening the door does not block it at the same time. Entrance doors require at least thirty inches of overhead protection outside to shelter a person entering or leaving in bad weather.

Round doorknobs on entrance and interior doors can be replaced without much trouble by levers, which are easier to operate. Choose

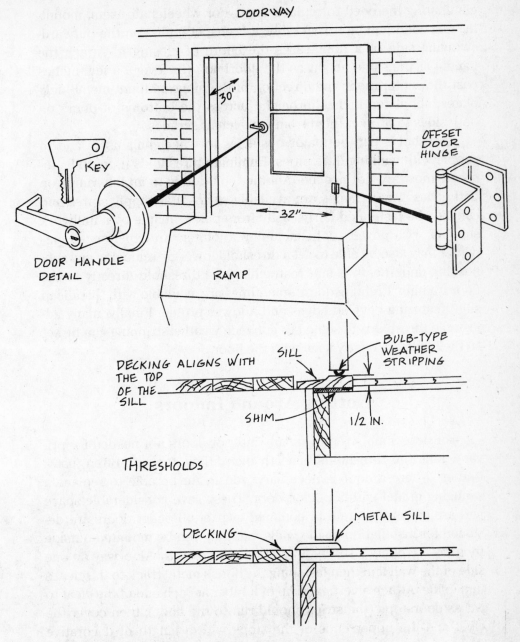

Illustration 112. Doorway accessibility features

levers whose ends curve back toward the wall; these do not catch on loose clothing. One patented lever fastens over existing doorknobs, eliminating the need to replace them. For wheelchair users, mounting a second handle—preferably, a D-shaped pull—on the outward-swinging side of a door eases the chore of closing it. Attach the handle at the same height as the doorknob (or lever), a few inches past the centerline in the direction of the hinges. Mount metal kick-plates, about ten inches in height, across the bottoms of doors on both sides to protect them from wheelchair scrapes.

Thresholds can be major obstacles to wheelchair users and to people with walking difficulties. Eliminate thresholds if possible; indoors, most are not needed. Another solution is to attach ramps on both sides. These ramps are available at hospital supply stores and can also be fashioned fairly easily from oak or maple by a millwork supplier with power tools, including a jointer. An experienced carpenter may also be able to set a threshold lower by removing any subflooring underneath it and reattaching the threshold directly to the floor framing. Prehung doors sometimes are available with "handicap sills" featuring beveled edges and a lower profile. Finally, many adjustable thresholds designed to enhance weatherstripping can be set so their sills are nearly level with the floor.

Getting Around Indoors

Consider simply removing any interior doors not needed for privacy. Like the thresholds beneath them, they, too, are often just a bother. Where a door is necessary, you might be able to replace a swinging model with a pocket door. These save considerable space and are easy to operate in confined quarters. Pocket doors are designed for installation inside walls, but they can be mounted outside them by attaching a sturdy horizontal cleat over the doorway on one side of the wall and then fastening the door's metal track to the cleat's underside. Attach a vertical strip of lumber at each end of the cleat to act as doorstops (the strips should run to the floor); then cover the cleat and the upper ends of the stops with a length of decorative molding. Use screws to attach the molding in case you need to remove it someday to repair the door. Also, replace the door's recessed handles with D-shaped handles or drawer pulls, which afford an easy grip.

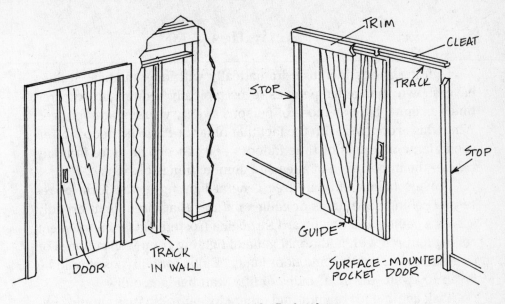

Illustration 113. Pocket doors

For a complete discussion of staircases and banisters, see the chapter "Avoiding Falls and Other Injuries." Short of installing an electric-powered lift, little can be done to modify stairs for disabled users except to make them safer. In some cases it may be possible to replace a staircase consisting of one to three steps (for example, steps leading to a sunken living room) with a ramp.

Floors are also discussed in "Avoiding Falls and Other Injuries." As was mentioned there, these must be firm and nonslippery—whether your concern is wheelchair users or persons with physical limitations. Replace thick or fraying carpet with tightly woven, foam-backed carpet or with vinyl flooring or tiles. Wood flooring is acceptable if it is not made slippery by waxing. Avoid using throw rugs.

When decorating, choose matte finishes for walls and countertops to reduce glare, and select solid-color (or at least simple) floor patterns that are not confusing to the eyes. Outfit furnishings, draperies, outlets, wall switches, and similar items in colors that contrast strongly with the walls. The purpose behind all of these strategies is to help alleviate the problems some older people have in distinguishing objects or in sensing differences in elevation and dimension.

Activities

Lighting needs increase dramatically with age; medical studies have shown that older peoples' eyes typically require up to three times as much light as those of persons twenty years old or younger. When this is coupled with the fact that older or disabled people often spend large amounts of time indoors—80 percent or more in many cases—the importance of adequate illumination is obvious.

Provide balanced, glare-free general lighting in the form of recessed ceiling downlights or ceiling fixtures that direct light broadly across a ceiling and downward through a frosted shade. Supplement ceiling fixtures with additional general lighting from wall sconces or track-mounted lighting and floor lamps. Bulbs for all types should be 60- to 100-watt "daylight" bulbs or fifty-watt halogen bulbs.

Task lighting for reading and other close-up activities should consist of 100- to 200-watt glare-free bulbs or comparable halogens. Install fluorescent lighting over a stove and under kitchen wall cabinets; in bathrooms, mount extensible illuminated vanity mirrors (the kind that magnify) on the wall near the basin. Be sure stairways are adequately illuminated; there should be switches for stairway lights at the top and bottom of the stairs, and the lighting itself should not cast confusing shadows that might be mistaken for steps. Install night-lights in pathways between bedrooms and bathrooms. Wherever possible, lighting of all kinds should be controlled by large wall switches, described below. Many disabled people cannot operate small switches on light fixtures.

Persons in wheelchairs can comfortably reach an area covering roughly twenty to forty-four inches above floor level. Therefore, lowering wall switches to about thirty-six inches from the floor and raising outlets to about thirty inches from the floor makes them more accessible both from a wheelchair and to people who have difficulty bending. Install a wall phone so the dial is slightly below eye level—forty-two to forty-four inches—for a seated person. Helpful, too, is replacing ordinary small toggle switches with large button controls, rockers, or keypads. Avoid switches with knobs that must be turned. Switches with internal illumination are easy to find in the dark. Heights of switches and outlets can be altered without cutting into

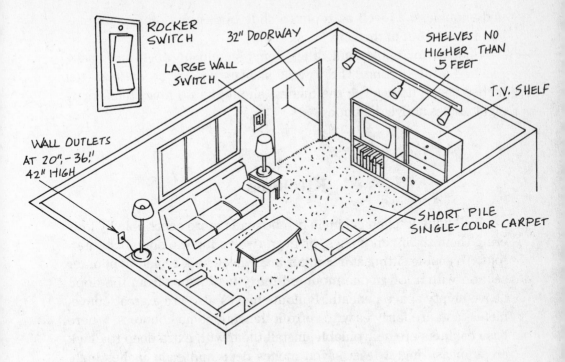

ROCKER SWITCH

32" DOORWAY

LARGE WALL SWITCH

SHELVES NO HIGHER THAN 5 FEET

T.V. SHELF

WALL OUTLETS AT 20" - 36" 42" HIGH

SHORT PILE SINGLE-COLOR CARPET

Illustration 114. An accessible living room

walls by using surface wiring fixtures linked to the existing switches and outlets.

Furniture in small rooms should be located around the perimeter to maintain space in the center for traffic flow. In large rooms, place furniture in the center to establish a four-to-five-foot path for traffic along the walls. Allow three to six feet between seating units. Sofas and chairs must be firm to provide support; recommended seat height is eighteen inches.

Desks for wheelchair users must be drop-front models or have wide spacing between the ends. The top surface should be about thirty-two inches above the floor and have a clearance of thirty inches underneath to accommodate the wheelchair's arms. A desk must measure about two feet front to back to cover a wheelchair oc-cupant's lap and legs.

Another style of desk or activity table to consider is a U-shaped, wraparound design. Such desks provide extra support for the arms and shoulders, as well as room to slide books and other materials comfortably out of the way.

Shelves for books and other items should be open and easily reached while sitting. The lowest shelves should be at least ten inches above floor level; the highest should be no more than forty-eight inches above the floor.

Kitchens

Wall-hung kitchen cabinets can be made more accessible by lowering them to fifteen inches, or even twelve inches, above countertops. To enable sitting at a counter in a wheelchair, replace a counter section with a lower one mounted thirty-two inches from the floor. Leave empty space beneath. Pullout shelves also are a great convenience and are fairly easy to retrofit beneath some counters. Where base cabinets are unavoidable, install them with extra-deep toe kick areas measuring at least seven inches deep and eight inches high. Leave the doors off or replace the latches with magnetic catches. Replace fixed cabinet shelves with pullout drawers, bins, and lazy Susans. Replace knobs on drawers and doors with D-shaped pulls.

The easiest mode of conventional cooking by an elderly person or a person in a wheelchair is over an electric cooktop installed in a dropped countertop that places the burners at a height of thirty-two inches above the floor (for wheelchair users, the space beneath the cooktop must be open). Look for a cooktop that has staggered burners (so the user does not have to reach directly across one burner to tend something on another) and plainly labeled, logically arranged controls conveniently mounted on the front. Touchpad burner controls are easier to use and to clean than knobs.

Conventional stove ovens are not recommended. Not only are the racks too low and awkward, the drop-front door is unmanageable for wheelchair users and can be dangerously hot as well. A better solution is a wall-mounted conventional or microwave oven with a side-mounted door, plenty of counter space near the latch side for setting down dishes, and a pullout shelf underneath that is high enough to clear a wheelchair user's knees. As is true for a cooktop, the base of either type of oven should be thirty-two inches from the floor.

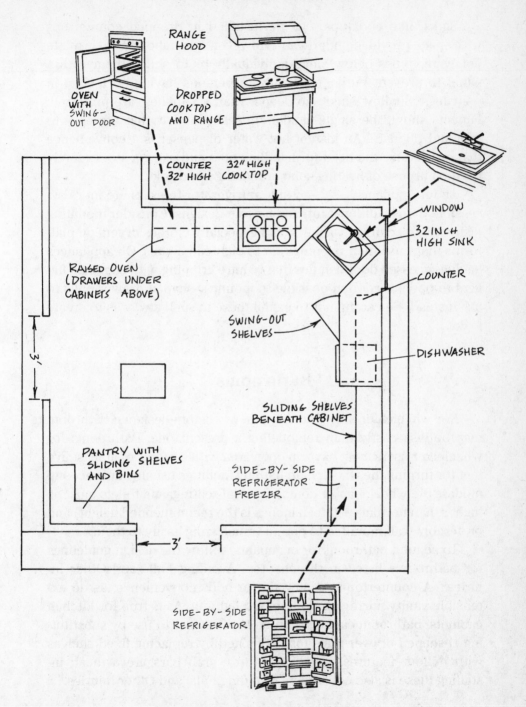

Illustration 115. An accessible kitchen

321

Sinks, like cooktops, can be installed in dropped countertops with space provided underneath. However, it is important to insulate hot water pipes below sinks individually or to shield them with a panel to prevent burns, especially to persons having no feeling in their legs. Shallow sinks, no deeper than six inches, are preferred. Faucets should be single-handle antiscald models that are easy to grip and regulate. An instant hot water dispenser is a convenience that can reduce the need for handling pots and using a stove merely for the purpose of heating soup or making hot drinks.

As for appliances, side-by-side refrigerator-freezers are more accessible than traditional top-and-bottom designs. Consider mounting dishwashers, clothes washers, and especially clothes dryers on platforms that raise their openings six to eighteen inches. On appliances having recessed door handles that require gripping with clenched fingers to open, it is often possible to mount D-shaped drawer pulls to use instead. Be careful not to install these in such a way as to create leaks.

Bathrooms

For wheelchair users, adding a new, custom-designed bathroom may be less costly than remodeling a present one. Bathrooms for wheelchair-users must have an open area with a radius of at least five feet for turning the wheelchair; bathing facilities usually must accommodate the wheelchair or contain special seating; and toilets must be higher than normal (eighteen inches is the recommended height) and preferably wall-mounted to permit transferring easily to them.

To adapt a bathroom for occupants, follow the design guidelines for bathrooms listed in the chapter "Avoiding Falls and Other Injuries." A countertop-to-ceiling mirror adds convenience, as do extensible vanity mirrors and increased lighting. As is true for kitchen cabinets, bathroom cabinets can be made easier to use by substituting D-shaped drawer pulls for knobs and by replacing fixed shelves with pullout features. Naturally, sturdy grab bars are a must; installing these is also covered in "Avoiding Falls and Other Injuries."

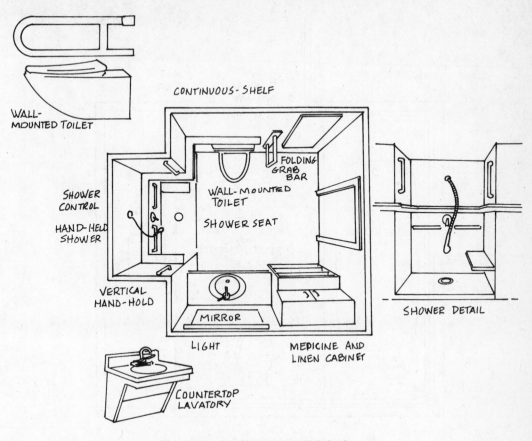

WALL-MOUNTED TOILET

CONTINUOUS-SHELF

FOLDING GRAB BAR

SHOWER CONTROL

HAND-HELD SHOWER

WALL-MOUNTED TOILET

SHOWER SEAT

VERTICAL HAND-HOLD

MIRROR

LIGHT

MEDICINE AND LINEN CABINET

SHOWER DETAIL

COUNTERTOP LAVATORY

Illustration 116. An accessible bathroom

Bedrooms

Ideally, a bedroom for an elderly or disabled person should be only a few easy steps from a bathroom. Privacy is important, as is uncluttered space, plenty of light and air, and an exit other than the main door.

If the door to a bedroom is in the middle of a wall, consider relocating it to a corner. This enables the entire wall to be used for storage or other purposes and usually affords privacy even when the door is open. Try to place the bed so there is at least thirty-six inches of clear space on both sides. Even if the occupant does not use a wheelchair, having plenty of room on each side of the bed simplifies getting in and out, doing things while seated on the bed, and changing

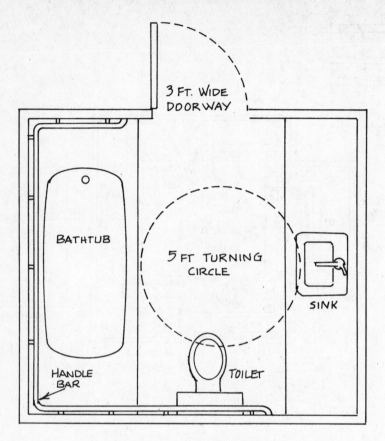

3 FT. WIDE
DOORWAY

BATHTUB

5 FT TURNING
CIRCLE

SINK

HANDLE
BAR

TOILET

Illustration 116A. Bathroom floor plan

bed linen. Provide ample nightstand arrangements and plenty of electrical outlet capacity. If possible, have an electrician install at least four outlets on each side of the bed, against the headwall so cords from appliances or medical equipment will be out of the way. Other outlets, including cable television hookups, should be installed or relocated from their existing positions with an eye to where they will be most convenient for someone in bed. Also install a three-way switch circuit that allows controlling bedroom lighting both from the doorway and from the bedside; this eliminates having to cross the room in the dark whether entering or leaving.

Bedroom closets should have sliding or pocket doors to maximize space. Inside a closet, clothes and other belongings should be arranged on lowered hanger poles and on shelves for easy access. Re-

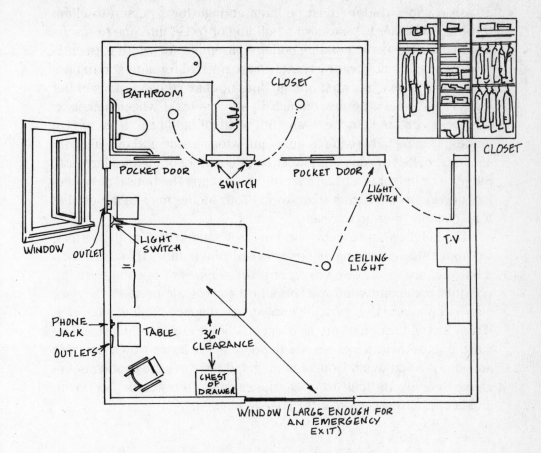

Illustration 117. An accessible bedroom

modeling a closet by adding homemade partitions to support such features is not too difficult; home centers and interior design stores also carry manufactured closet storage systems that are easy to customize and assemble. Adequate lighting inside a closet is important, but consult a building inspector before adding any; strict requirements generally apply to insure that closet lights are not fire hazards.

Many building codes now require two exits from each room in a house, to permit emergency escape in case of fire. This regulation typically applies only to new construction (including additions), but remodeling existing rooms to provide emergency egress should be a renovation priority in any house—and is a vital necessity in a bedroom occupied by someone who is disabled.

An escape window must be large enough for a person to climb through, whether to get out of a building or to get into one to rescue occupants. The usual building code requirement for width is twenty-four inches, but at least thirty inches are needed for a person to pass who cannot move headfirst or who does not have the use of his or her legs. The sill must be low enough to sit on—for a wheelchair user, this means eighteen inches—and the width should be at least twelve inches. This usually requires installing wider inside and outside trim sections, called aprons, that must be adequately braced underneath. Such a low height is a hazard if children occupy the house; nevertheless, a window sill higher than twenty-four inches may be impossible for a disabled person to reach.

The height of the escape window itself should be four to five feet. On upper stories, sliding glass doors or French doors that open onto a balcony are ideal. Elsewhere, or if these types of openings are impossible, casement windows serve best for escape because they are easy to open and they provide unobstructed space. Sliding windows, if they are wide enough, are also a good choice and should be used instead of casement windows where opening the latter would obstruct an outdoor walkway. Double-hung windows present problems because they are difficult to open, the sashes often jam or fall down, and because they seldom afford needed height.

Appendix

Electromagnetic Fields

VIRTUALLY EVERY ELECTRICAL SOURCE GENERATES AN ELECTROMAGNETIC field (EMF). Technically, EMFs consist of two parts, an electric field that is shielded by normal wiring insulation, and a magnetic field that passes through almost any material, including human tissue. Scientists are studying the effects of magnetic fields on human cells. More than forty studies worldwide during the past twenty-five years have associated EMFs with some types of cancer, but none has proven that EMFs cause or promote cancer in humans. While it may be prudent to avoid exposure to electromagnetic fields whenever possible, eliminating them from a house is impossible and arguably overzealous.

The earth itself produces an enormous magnetic field that is relatively unchanging and permeates everything. The electrical current produced by most power companies is alternating current (AC); the magnetic field it generates is much smaller and changes by reversing direction twice with each electrical cycle. Normal sixty-cycle household current, known as 60-Hertz (Hz), completes a cycle sixty times per second; during that time it generates 120 magnetic pulses.

Most American homes are believed to contain a background level of electromagnetic radiation measuring between 0.1 and 2 milligauss (mG)—a gauss is the standard measuring unit for electromagnetism—according to studies performed by various private and government-funded groups during the 1970s and 1980s. Medical research has focused on fields measuring from 2 to 3 mG.

The strongest emitters of EMFs are electric transmission lines, which carry power at voltages often exceeding 500,000 volts from generating stations to smaller distribution lines (for more on how electricity reaches your home, see the chapter "Upgrading Old Wiring"). Such strong electrical currents undeniably generate strong magnetic fields; however, the strength of a magnetic field drops off rapidly with distance from its source. At a distance of one hundred yards from even the most powerful transmission line (generating EMFs of over 100 mG), the magnetic field measurement may be as little as 1 mG—normal background level. Neighborhood power lines offer somewhat more threat because they run closer to houses; EMFs from these can be in the area of 2 mG or higher. Another source of EMFs is faulty household grounding systems (these, too, are discussed in "Upgrading Old Wiring").

Most contemporary household wiring generates little or no magnetic field. This is because branch circuit cables typically consist of two wires—one hot and one neutral—that carry current in opposite directions. Because the wires, or conductors, are in close proximity, their fields tend to cancel each other. However, old-fashioned knob-and-tube wiring consists of conductors separated widely enough that substantial magnetic fields can exist. Powerful enough to deflect a compass needle—one test that the wiring carries current—EMFs from knob-and-tube circuits have been measured at 7 mG a few feet away, according to *Consumer Reports*.

Some household appliances and electronic equipment produce significant EMFs. Also according to *Consumer Reports*, these include electric blankets (20 mG or more), microwave ovens (10 mG or more), televisions (60 mG or more at the sides), and computers (computer monitors have approximately the same levels as televisions). In the case of electric blankets, it is recommended that only newer models advertising low EMFs be used or that extra blankets or a quilt be substituted. With the other items, keeping them at arm's length (two to four feet) during use minimizes contact with their magnetic fields, which at that distance are around 2 mG or less.

Gauss meters for measuring EMFs are available from electronics stores. So far, however, models for home use generally have not been found reliable. To have EMFs measured in or around your home, your best bet probably is to call your electric utility company. It might even perform the service free of charge.

Seasonal Affective Disorder

Winter months, especially in northern latitudes, have long been associated with gloomy moods, lethargy, social withdrawal, and even death (numerous studies have shown suicide rates in northern regions to be generally higher in winter than in summer). But it has only been during the last ten or fifteen years that these symptoms of depression that characteristically appear in late fall and vanish around the time of the vernal equinox have been identified by the medical community as an actual malady, now termed Seasonal Affective Disorder (SAD).

Perhaps because public awareness of SAD has developed during roughly the same time as the raising of public consciousness concerning environmental issues and indoor air quality, the two occasionally are popularly confused as being at least contextually linked, if not somehow directly related.

However, SAD has been demonstrated to result primarily from a deficiency of light (perhaps influencing the pineal gland), not from pollutants in the atmosphere. The most common and successful form of treatment is simply administering moderate doses of high-intensity light—up to twenty times the strength of normal indoor lighting—once or twice daily to patients. Typically, the condition then disappears within one to four weeks.

From the standpoint of remodeling a home to avoid the onset of SAD or to counter its effects, an obvious strategy is to provide as much natural light by means of windows and skylights as possible, and to have these light sources located where they will capture the sun's direct rays during the months when days are shortest.

Installing plenty of artificial light should help, too; fortunately, there seems to be no loss of benefit whether the artificial light is incandescent, fluorescent, or so-called full-spectrum. Full-spectrum lighting is a fluorescent type that more closely duplicates the rays of natural sunlight than do either incandescent or ordinary fluorescent bulbs. It has been criticized for emitting excess ultraviolet radiation, which can cause serious health effects—including eye damage and cancer. For that reason, using full-spectrum lights regularly and without adequate UV screening is not advised, and the U.S. Food and Drug Administration has not approved them as safe and effective for treating SAD.

If you suspect that you or other household members suffer from SAD, see a doctor.

Multiple Chemical Sensitivity

Pollutants, whether man-made or natural, are not breathed individually. Rather, they enter the body in complex, changing mixtures. So far, research has yet to explore, let alone uncover, the health effects of blended pollutants, but there is a growing feeling among some doctors that such effects do exist and that a segment of the population suffers from them in a variety of ways. Indeed, according to a study commissioned by the New Jersey Department of Health, published in 1990, annual chemical production in the United States has grown from 8 million tons right after World War II to 108 million tons in 1985. It is probably unreasonable to think that no health effects are to be expected from this massive dumping of potentially harmful compounds into the environment; on the other hand, none have been conclusively demonstrated.

Symptoms of multiple chemical sensitivity, or MCS (other names for this illness are chemical hypersensitivity, total allergy syndrome, twentieth-century disease, and environmental illness), range from those typical of allergic reactions (nasal and upper respiratory congestion, watery eyes, rashes, and asthma) to headaches, dizziness, fatigue, nausea, depression, and confusion. Linking cases, according to the doctors who treat them, is their occurrence in affected individuals at lower and lower chemical exposures and their provocation by an ever-broadening variety of compounds. The belief is that single or multiple exposures to certain chemicals can cause increased sensitivity to many more substances, often at infinitesimal levels. Some specialists speculate that patients—and probably all living things—possess an overall limit to stress in general, and that once the "total body load" of stress from individual and combined chemical, biological, psychological, and physical sources passes a certain critical threshold, even very low levels of additional stress from any source can produce allergic response.

Certainly it is the purpose of *The Healthy Home Handbook* to address the presence of potentially harmful chemicals in homes, with the aim of eliminating them as hazards. But for people suffering from MCS in all but its mildest forms it is likely that the measures outlined in these pages will not be enough. Even ordinary allergies should be viewed first as a medical problem, and any household modifications undertaken to alleviate them should be performed as adjunct activities to medical advice and treatment.

Modifying and building homes for chemically sensitive occupants is possible and has been done successfully by a growing number of architects and builders, many of whom now specialize in the process. In the main, it involves putting to use all of the information contained in this book, but to a greater degree and with a higher standard of meticulousness. Obviously, creating a home that is totally free of irritants can be expensive—not to mention complicated, given the differences in each person's response to the environment. Before pursuing the goal of an allergy-free home, spend time studying and planning extensively by using the latest sources and the advice of experts with proven successes in this field. Otherwise the result could be financial fiasco and disappointment, with no relief from symptoms.

Where to Find Help

THE FOLLOWING LIST OF ORGANIZATIONS, MANUFACTURERS, SUPPLIERS, books, and other publications can help you locate additional information and products on household health hazards. Naturally, the list is by no means complete and is intended only as a resource; mention does not imply endorsement, and readers are advised to compare products and services thoroughly before purchasing. Many manufacturers and suppliers also offer free informative literature and catalogs.

Newsstand magazines for homeowners and general consumers are an excellent source of the latest information. Among these are *The Family Handyman* (7900 International Drive, Suite 950, Minneapolis, MN 55425; 800-285-4961), *Home Mechanix* (2 Park Avenue, New York, NY 10016; 800-456-6369), *Fine Homebuilding* (63 South Main Street, P.O. Box 5506, Newtown, CT 06470; 800-888-8286), *The Journal of Light Construction* (RR 2, Box 146, Richmond, VT 05477; 800-375-5981), *This Old House* (20 West 43rd Street, New York, NY 10036; 212-522-9465) and *Consumer Reports* (101 Truman Avenue, Yonkers, NY 10703). Other sources of current information are home- and health-related forums on computer on-line services such as CompuServ, America Online and the Microsoft Network, and the home pages of product manufacturers, associations, and government organizations on the World Wide Web.

When searching for books, don't overlook *Subject Guide to Books in Print*, a directory of current books arranged by subject, published every two years by R. R. Bowker, New Providence, NJ. Companion volumes list books arranged by title and author. You'll find these a standard reference in virtually all public libraries.

General

American Lung Association
1740 Broadway
New York, NY 10019
(212) 315-8700
(800) 586-4872
Free booklets and research advice on
all types of indoor air pollutants. Offices
are located in every state.

Bonneville Power Administration
Box 12999
Portland, OR 97212
(503) 230-3478
Information on energy conservation and
indoor pollution levels

331

*Consumer Product Safety
Commission*
5401 Westbard Avenue
Washington, DC 20207
(800) 638-2772
Free booklets and information on product and household safety

*Department of Housing and Urban
Development*
451 7th Street, S.W.
Washington, DC 20410
(202) 755-6420
Information on Minimum Property Standards and HUD regulations pertaining
to pollutants including lead, formaldehyde, and noise. For publications list
and requests, contact HUD User's Document Reproduction Service, P.O. Box
6091, Rockville, MD 20850; (800) 245-2691 or (301) 251-5154.

*National Association of Home
Builders*
Energy and Home Environment Department
15th and M Streets, N.W.
Washington, DC 20005
(800) 368-5242
Information about indoor air quality and
building products. For publications list
and ordering information, contact
NAHB Bookstore; (800) 368-5242.

National Safety Council
P.O. Box 33435
Washington, DC 20033
(800) 767-7236
Voluntary nongovernmental organization. Free newsletters and publications
on accident-prevention methods

*Occupational Safety and Health
Administration*
200 Constitution Avenue, N.W.
Washington, DC 20210
(800) 582-1708
(202) 523-8148 (in Washington, DC)

Free booklets and information on workplace safety and health standards

*U.S. Environmental Protection
Agency*
401 M Street, S.W.
Washington, DC 20460
(202) 829-3535 (public information center)
(202) 382-4355 (EPA publications requests)
Free booklets, publications, and information on indoor air quality subjects

EPA Regional Offices

Region 1 (Connecticut, Maine,
Massachusetts, New Hampshire,
Rhode Island, Vermont)
(617) 565-3420

Region 2 (New Jersey, New York,
Puerto Rico, Virgin Islands)
(212) 637-3000

Region 3 (Delaware, Washington,
DC, Maryland, Pennsylvania, Virginia, West Virginia)
(215) 597-9800

Region 4 (Alabama, Florida, Georgia, Kentucky, Mississippi, North
Carolina, South Carolina)
(404) 347-4727

Region 5 (Illinois, Indiana, Michigan, Minnesota, Ohio, Wisconsin)
(312) 353-2000

Region 6 (Arkansas, Louisiana,
New Mexico, Oklahoma, Texas)
(214) 665-6444

Region 7 (Iowa, Kansas, Missouri,
Nebraska)
(913) 551-7000

Region 8 (Colorado, Montana,
North Dakota, South Dakota, Utah,
Wyoming)
(303) 293-1603

Region 9 (Arizona, California,
Hawaii, Nevada)
(415) 744-1305

Region 10 (Alaska, Idaho, Oregon,
Washington)
(206) 553-1200

Books and Publications

Directory of State Indoor Air Contacts. Public Health Foundation, 1220 L Street, N.W., Washington, DC 20005; (202) 898-5600. Free booklet listing state agency contacts for individual air quality issues, regional offices of EPA and OSHA, and national hot lines for EPA, CSPC, and the U.S. Department of Health and Human Services.

Sweet's Catalog Files. Published annually by Sweet's Group, McGraw-Hill Inc., 1221 Avenue of the Americas, New York, NY 10020; (800) 992-0187. Current product literature from a wide range of manufacturers; separate volumes for general building and renovation, home building and remodeling, accessible building products, lighting, kitchen and bath remodeling, and interior decoration.

The Thomas Register of American Manufacturers. Published annually by Thomas Publishing Company, 5 Penn Plaza, New York, NY 10001; (212) 695-0500. Directory of building materials and products manufacturers.

Asbestos

Organizations

American Lung Association
GPO Box 596
New York, NY 10116-0596
(800) 586-4872

Consumer Product Safety Commission
Washington, DC 20207
(800) 638-2772

National Asbestos Council/The Environmental Information Association
1777 Northeast Expressway, Suite 150
Atlanta, GA 30329
(404) 633-2622
(404) 633-5714 (fax)

National Institute of Standards and Technology
B321 Research Building
Gaithersburg, MD 20899-0001
(301) 975-4016

Resilient Floor Covering Institute
966 Hungerford Drive, Suite 12B
Rockville, MD 20850
(301) 340-8580
(301) 340-7283 (fax)

U.S. Environmental Protection Agency
Toxic Substances Control Act (TSCA)
(202) 554-1404
(800) 835-6700 (asbestos information hot line)

Product Manufacturers and Suppliers

*Eastern Safety Equipment
Company, Inc.*
59-20 56th Avenue
Maspeth, NY 11378
(718) 894-7900
Dust masks, respirators, protective
clothing, and other safety equipment

E. I. Dupont DeNemours and Co.
P.O. box 80705
Wilmington, DE 19880
(800) 448-9835 (800/44-TYVEK)
Protective clothing

3M Filtration Products
76-1 W, 3M Center
St. Paul, MN 55144
(800) 383-6288
(612) 733-1110
(800) 666-6477 (sales assistance)
(800) 243-4630 (technical service)
Air filters, dust masks, and respirators

*International Protective Coatings
Corp.*
725 Carol Avenue
Ocean Township, NJ 07755
(800) 334-8796
(908) 531-5192 (fax)
Asbestos-abatement materials and coatings

Lab Safety Supply, Inc.
P.O. Box 1368
Janesville, WI 53547-1368
(800)356-0783
Dust masks, respirators, protective
clothing, and other safety equipment

Mine Safety Appliances Company
P.O. Box 426
Pittsburg, PA 15230
(800) 672-2222
Dust masks, respirators, protective
clothing, and other safety equipment

Nilfisk of America
300 Technology Drive
Malvern, PA 19355
(215) 647-6420
(215) 647-6427 (fax)
Dust-collection systems and equipment,
including HEPA vacuums and filters

Northern Safety Company
P.O. Box 4250
Utica, NY 13505
(800) 631-1246
Dust masks, protective clothing, and
other safety equipment

Books and Publications

Asbestos in the Home. 1986. U.S. Consumer Product Safety Commission. Available from regional EPA offices and from Superintendent of Documents, U.S. Government Printing Office, Washington, DC 20402. Pamphlet outlining basic residential asbestos issues, precautions, and remedies.

Abatement of Asbestos-Containing Pipe Insulation. 1986. Technical Bulletin No. 1986-2. U.S. Environmental Protection Agency. Available from regional EPA offices and from TOSCA Assistance Hot Line; (202) 554-1404. Technical bulletin outlining procedures for encapsulating and removing asbestos from pipes.

Guidance for Controlling Asbestos-Containing Materials in Buildings (EPA 560/5-85-024). 1985. U.S. Environmental Protection Agency. Available from regional EPA offices and from TOSCA Assistance Hot Line; (202) 554-1404. Detailed handbook for addressing problems of asbestos in residential and nonresidential buildings.

Lead

Organizations

American Water Resources Association
5410 Grosvenor Lane, Suite 220
Bethesda, MD 20814
(301) 493-8600

Clean Water Fund
317 Pennsylvania Avenue, S.E., Suite 200
Washington, DC 20003
(202) 547-2312

National Lead Abatement Council
P.O. Box 535
Olney, MD 20832
(800) 673-8202
(301) 924-5490 (in Maryland)
(301) 924-0265 (fax)

U.S Department of Health and Human Services
Public Health Services
Centers for Disease Control
Atlanta, GA 30333
(404) 488-7330

U.S Department of Housing and Urban Development
451 7th Street, S.W.
Washington, DC 20410

(202) 755-6420
For publications list and requests, contact HUD User's Document Reproduction Service, P.O. Box 6091, Rockville, MD 20850; (800) 245-2691 or (301) 251-5154.

U.S. Environmental Protection Agency
401 M Street, S.W.
Washington, DC 20460
(202) 829-3535 (public information center)
(202) 382-4355 (EPA publications requests)
(800) 532-3394 (National Lead Information Center)
(800) 424-5323 (National Lead Immigration Clearinghouse)
(800) 426-4791 (Safe Drinking Water Hot Line)
Also contact EPA regional offices.

Water Quality Association
4151 Naperville Road
Lisle, IL 60532
(708) 505-0160

Product Manufacturers and Suppliers

Acc-U-Test
P.O. Box 143
East Weymouth, MA 02189
(617) 337-5546
Acc-u-Test (brand) sodium sulfide test kits

Carolina Environment
P.O. Box 2661
Charlotte, NC 28221
(800) 448-5323
Know Lead (brand) rhodizonate paint-testing kits

Casto Enterprises
P.O. Box 39
Kenna, WV 25248
(304) 372-4237
No Swett (brand) lead-base paint stripper

Clean Water Lead Testing, Inc.
1 University Heights
Asheville, NC 28804
(704) 251-6800
Mail-in testing kits for tap water

Culligan
One Culligan Parkway
Northbrook, IL 60062
(800) 285-5442 (800-CULLIGAN)
Water filters

Desco Manufacturing Company, Inc.
1445 West Cowles Street
Long Beach, CA 90813
(800) 337-2648
(310) 436-0280 (fax)
Dust-containing tools and HEPA vacuums

Dumond Chemicals, Inc.
1501 Broadway
New York, NY 10036
(212) 398-0815 (fax)
Peel Away (brand) poultice-type paint-removal systems

Dynacraft Industries
4 Kinney Road
Englishtown, NJ 07726
(908) 303-8920
(908) 409-1144 (fax)
Encapsulants; paint-removal chemicals

Eagle Industries
P.O. Box 10652
New Orleans, LA 70181
(504) 733-3510
(504) 733-3552
Containment, ventilation, and decontamination equipment

Eastern Safety Equipment Company, Inc.
59-20 56th Avenue
Maspeth, NY 11378
(718) 894-7900
Dust masks, respirators, protective clothing, and other safety equipment

E.I. Du Pont DeNemours and Co.
P.O. Box 80705
Wilmington, DE 19880
(800) 448-9835 (800/44-TYVEK)
Protective clothing

Elasti-Cote, Inc.
34 South Market Street
Frederick, MD 21701
(301) 663-0411
(301) 663-0448
Encapsulants

Encap Systems Corporation
230 North Central Avenue
Columbus, OH 43222
(614) 274-3666
(614) 274-3444 (fax)
Encapsulants

Encapsulation Technologies Corp.
310-312 North Charles Street
Baltimore, MD 21201-4302
(301) 962-5335
Encapsulants

Enzone, Inc.
P.O. Box 290480
Davie, FL 33329-0480
(800) 448-0535
Lead Zone (brand) rhodizonate test kits

Euroclean
905 West Irving Park Road
Itasca, IL 60143
(708) 773-2111
(708) 773-2859 (fax)
HEPA vacuums

Everpure
660 Blackhawk Drive
Westmont, IL 60559
(708) 654-4000
Water filters

FibreWall
2546 West Seventh Street
Los Angeles, CA 90057
(213) 385-4800
Woven and textured fiberglass wall coverings

Flexi-Wall Systems
P.O. Box 88
Liberty, SC 29657
(800) 843-5394
(800) 843-9318 (fax)
Gypsum-saturated fabric wall covering

3G Mermet Corporation
3963 Virginia Avenue
Cincinnati, OH 45227
(800) 847-7243
Rustiver (brand) woven and textured fiberglass wall coverings

The Glidden Company
925 Euclid Avenue
Cleveland, OH 44115
(216) 344-8900
Glid-Wall (brand) fiberglass wall coverings

Hybrivet Systems, Inc.
P.O. Box 1210
Framingham, MA 01701
(800) 262-5323
(800) 242-4325
LeadCheck Swabs (brand) rhodizonate testing kits for paint, LeadCheck Aqua (brand) testing kits for tap water

Hin-Cor Industries, Inc.
136 Sea Island Parkway, Suite 10502
Beaufort, SC 29902
(803) 522-3066
(803) 522-8470 (fax)

Ledizolv (brand) phosphate-free lead-dissolving detergent

Innovative Syntheses Corporation
2143 Commonwealth Avenue
Newton, MA 02166
(617) 965-5653
Lead Detective (brand) sodium sulfide paint-testing kit

International Protective Coatings Corp.
725 Carol Avenue
Oakhurst, NJ 07755
(800) 334-8796
(908) 531-5192 (fax)
Lead-abatement products, including paint strippers and coatings

Lab Safety Supply, Inc.
P.O. Box 1368
Janesville, WI 53547-1368
(800) 356-0783
Dust masks, protective clothing, and other safety equipment

Lead-Free Faucets, Inc.
16715 West Park Circle Drive
Chagrin Falls, OH 44022
(216) 543-1660
Lead-free faucets

3M Filtration Products
76-1W, 3M Center
St. Paul, MN 55144
(800) 383-6288
(612) 733-1110
(800) 666-6477 (sales assistance)
(800) 243-4630 (technical service)
Air filters, dust masks, and respirators

Mine Safety Appliances Company
P.O. Box 426
Pittsburgh, PA 15230
(800) 672-2222
Dust masks, respirators, protective clothing, and other safety equipment

Moldex-Metric, Inc.
10111 West Jefferson Boulevard
Culver City, CA 90230
(310) 837-6500
(310) 837-9563
Respiratory protection

Neoterik Health Technologies, Inc.
401 South Main Street
Woodsboro, MD 21798
(800) 895-4880
(301) 845-2777
(301) 845-2213
Respirators

Nikro
638 North Iowa Street
Villa Park, IL 60181
(800) 875-NIKRO
HEPA vacuums and filters

Nilfisk of America
300 Technology Drive
Malvern, PA 19355
(215) 647-6420
(215) 647-6427 (fax)

Dust-collection systems and equipment,
including HEPA vacuums and filters

Northern Safety Company
P.O. Box 4250
Utica, NY 13505
(800) 631-1246
Dust masks, protective clothing, and
other safety equipment

Purisys, Inc.
424 Madison Avenue
New York, NY 10017
(212) 308-1222
Home Diagnostics (brand) mail-in paint-
testing kit

Sensidyne
16333 Bay Vista Drive
Clearwater, FL 34620
(800) 451-9444 in Florida
(813) 530-3602
Lead Alert (brand) testing kits and other
detection equipment

Books and Publications

Comprehensive and Workable Plan for the Abatement of Lead-Based Paint in Privately Owned Housing. 1990. Available from HUD User's Document Reproduction Service, P.O. Box 6091, Rockville, MD 20850; (800) 245-2691 or (301) 251-5154. Report to Congress focusing on health effects of household lead and abatement procedures.

Historic Buildings and the Lead Paint Hazard. Available from Massachusetts Historical Commission, 80 Boylston Street, Boston, MA 02116; (617) 727-8470. Helpful booklet for owners or restorers of older homes.

Lead-Based Paint: Interim Guidelines for Hazard Identification and Abatement in Public and Indian Housing. 1990. Available from HUD User's Document Reproduction Service, P.O. Box 6091, Rockville, MD 20850; (800) 245-2691 or (301) 251-5154. Compendium of information on the need for and appropriate methods of identifying and abating lead-based paint; includes exhaustive appendices and tables.

Preventing Lead Poisoning in Young Children. 1991. Available from EPA regional offices and by telephoning EPA publications request number: (202) 382-4355. A statement by the Centers for Disease Control; emphasis on hazards of household lead in children and details of recommended abatement procedures.

What Remodelers Need to Know and Do About Lead. 1993. Available from National Association of Home Builders Bookstore, 15th and M Streets, N.W., Washington, DC 20005; (800) 368-5242. A comprehensive guide for professional remodelers and painters, including information on legal issues.

Radon

Organizations

American Lung Association
GPO Box 596
New York, NY 10116-0596
(800) LUNG-USA (586-4872)
Booklets on radon and other indoor pollutants

Indoor Air Division
Office of Air and Radiation
U.S. Environmental Protection Agency
(EPA)

401 M Street, S.W.
Washington, DC 20460
(800) 767-7236 (800-SOS-RADON) (EPA radon hot line)

National Safety Council
P.O. Box 33435
Washington, DC 20033
(800) 767-7236

Product Manufacturers and Suppliers

Air Chek, Inc.
570 Butler Bridge Road
Fletcher, NC 28732
(800) 247-2435
(704) 684-0893
Air Chek (brand) short-term test kits

BRK Brands, Inc.
3901 Liberty Street Road
Aurora, IL 60504
(800) 323-9005
First Alert (brand) short-term test kits

DSK Environmental Safety Products Inc.
325 North Oakhurst Drive
Beverly Hills, CA 90210
(213) 550-7600
DSK Safer Home Test Kit (brand), consisting of five separate tests for radon (short-term test), lead, microwave oven radiation, carbon monoxide, and ultraviolet radiation

Eastern Safety Equipment Company, Inc.
59-20 56th Avenue
Maspeth, NY 11378
(718) 894-7900
Dust masks, respirators, protective clothing, and other safety equipment

Key Technology
P.O. Box 562
Jonestown, PA 17038
(800) 523-4964
Key-Rad (brand) short-term test kits, Key-Rad long-term test kits

Landauer, Inc.
2 Science Road
Glenwood, IL 60425
(708) 755-7911
RadTrak (brand) long-term test kits

Lab Safety Supply, Inc.
P.O. Box 1368
Janesville, WI 53547-1368
(800) 356-0783
Dust masks, protective clothing, and other safety equipment

3M Filtration Products
76-1W, 3M Center
St. Paul, MN 55144
(800) 383-6288
(612) 733-1110
(800) 666-6477 (sales assistance)
(800) 243-4630 (technical service)
Air filters, dust masks, and respirators

Mine Safety Appliances Company
P.O. Box 426
Pittsburgh, PA 15230
(800) 672-2222
Dust masks, respirators, protective clothing, and other safety equipment

Northern Safety Company
P.O. Box 4250
Utica, NY 13505
(800) 631-1246
Dust masks, protective clothing, and other safety equipment

North East Environmental Products
17 Technology Drive
West Lebanon, NH 03784
(603) 298-7061
(603) 298-7063 (fax)
ShallowTray (brand) low-profile air stripper water-purification systems

Purisys, Inc.
424 Madison Avenue
New York, NY 10017

(212) 308-1222
Home Diagnostics (brand) short-term test kits

Radon Control Inc.
511 Industrial Drive
Carmel, IN 46032
(317) 846-7486
(317) 846-5882
Radon-mitigation equipment and supplies

Radon Reduction Supply Co.
P.O. Box 113
Hawleyville, CT 06440-0113
(203) 270-0803
Continuous radon-monitoring devices

RTCA
The Trent Building
P.O. Box 258
Irvington, NY 10533
RTCA (brand) short-term test kits

Safe-Aire, Inc.
P.O. Box 160
Canton, IL 61520
(309) 647-0331
Radon testing and mitigation products

Sun Nuclear Corp.
425A Pineda Court
Melbourne, FL 32940
(407) 259-6862
Continuous radon-monitoring devices

Teledyne-Brown Engineering, Inc.
50 Van Buren Avenue
P.O. Box 1235
Westwood, NJ 07675
(800) 666-0222
Teledyne (brand) short-term test kits

Books and Publications

A Citizen's Guide to Radon: The Guide to Protecting Yourself and Your Family from Radon (ANR-464). 1992. Available by calling the EPA radon hot line, (800) 767-7236; state office of radiation control; or EPA regional office. Concise, easy-to-understand overview of radon, its hazards, and how to minimize its risks.

Radon: The Invisible Threat. Michael Lafavore. 1987. Published by Rodale Press, Emmaus, PA. Popular guide to radon, its hazards, and mitigation.

Radon-Resistant Construction Techniques for New Residential Construction (EPA/625/2-91/032). 1991. Available from EPA by calling the EPA radon hot line, (800) 767-7236; state office of radiation control; or EPA regional office. Similar to *Radon-Reduction Techniques for Detached Houses* (EPA/625/5-87-019), but focuses on mitigation methods for new construction.

Radon-Reduction Methods: A Homeowner's Guide (RD-681). 1989. Available from EPA by calling the EPA radon hot line, (800) 767-7236; state office of radiation control; or EPA regional office. Concise, readable guide covering household radon detection and mitigation methods in moderate detail.

Radon-Reduction Techniques for Detached Houses: Technical Guidance (EPA/625/5-87-019). 1988. Available from EPA by calling the EPA radon hot line, (800) 767-7236; state office of radiation control; or EPA regional office. Professional's guide to radon detection and mitigation, covering methods in greatest detail.

Technical Support Document for the 1992 Citizen's Guide to Radon (EPA/400-R-92-011). 1992. Companion volume to *A Citizen's Guide to Radon* (ANR-464), containing exhaustive information upon which *A Citizen's Guide* is based.

Combustion Products

Organizations

American Gas Association
1515 Wilson Boulevard
Arlington, VA 22209
(703) 841-8400

American Lung Association
1740 Broadway
New York, NY 10019
(212) 315-8700
(800) 586-4872

Carbon Monoxide Safety and Health Association
11211 Sorrento Valley Road, Suite D
San Diego, CA 92121
(800) 432-5599

Chimney Sweep Institute of America/National Chimney Sweep Guild
16021 Industrial Drive, Suite 8
Gaithersburg, MD 20877

(301) 963-6900 (Institute)
(301) 963-6900 (Guild)

Hearth Products Association
1101 Connecticut Avenue, N.W., Suite 700
Washington, DC 20036
(202) 857-1181

Indoor Air Division
Office of Air and Radiation
U.S. Environmental Protection Agency
401 M Street, S.W.
Washington, DC 20460
(202) 260-7400

Masonry Fireplace and Chimney Association
Box 19264
Alexandria, VA 22320
(703) 549-2568

National Fire Protection
Association
1 Battery March Park
Quincy, MA 02269
(617) 770-3000

Product Manufacturers and Suppliers

ADDM International
P.O. Box 572
Oceanside, NY 11572
(516) 766-5997
Pama (brand) CO detectors

Ahrens Chimney Technique
2000 Industrial Avenue
Sioux Falls, SD 57104
(800) 843-4417
Ceramu-Flue (brand) high-quality ceramic chimney liners

Alladin Steel Products
401 North Wynne Street
Colville, WA 99114
(509) 684-3745
Quadra-Fire (brand) pellet stoves; improved woodstoves

American Sensors, Inc.
100 Tempo Avenue
North York, Ontario M2H 2N8
Canada
American Sensors (brand) CO detectors

Austroflamm
2210 Alexander Street, Suite A
Salt Lake City, UT 84119
(800) 443-9399
(801) 972-9400 (in Utah)
Improved woodstoves

Beechwood Industries
889 Horan Drive
Fenton, MO 63026
(314) 343-4100
Airtight fireplace doors

BRK Brands, Inc.
3901 Liberty Street Road
Aurora, IL 60504
(800) 323-9005
First Alert (brand) CO detectors

Chimtek
P.O. Box 233
Sparta, MO 65753
(417) 278-3320
Masonry chimney liners

Copperfield Chimney Supply
304 South 20th Street
Fairfield, IA 52556
(515) 472-4126
Metal chimney liners

Diamond Products
30 Railroad Avenue
Albany, NY 12205
(518) 459-6775
Airtight fireplace doors

The Earth Stove
10595 Southwest Manhasset
Tualatin, OR 97062
(503) 692-3991
Earth Stove (brand) pellet stoves

Enzone, Inc.
P.O. Box 290480
Davie, FL 33329-0480
(800) 448-0535
Enzone (brand) CO detectors

Golden/Flue
Route 3, Box 237
Rutherglen, VA 22546
(804) 798-1089
Cast masonry chimney liners

Goodway Tools Corporation
420 West Avenue
Stamford, CT 06902-6384
(800) 333-7467
(203) 359-4708
Fireplace-cleaning vacuum attachments

Heat Fab
38 Haywood Street
Greenfield, MA 01301
(800) 772-0739
(413) 774-2356 (in Massachusetts)
Metal chimney liners

Heatilator Inc.
1915 West Saunders Street
Mt. Pleasant, IA 52641
(800) 843-2848
Zero-clearance fireplaces

Heat-N-Glo
666 West Highway 13
Savage, MN 55378
(612) 890-8367
Zero-clearance fireplaces

Hearth-craft
116 South 10th Street
Louisville, KY 40402
(502) 589-6220
Airtight fireplace doors

Jameson Home Products
2820 Thatcher Road
Downers Grove, IL 60515
(800) 779-1719
Jameson (brand) CO detectors

Majestic
1000 East Market Street
Huntington, IN 46750-2579
(219) 356-8000
Zero-clearance fireplaces

Metal-Fab
P.O. Box 1138
Wichita, KS 67201
(800) 835-2830

(316) 943-2351 (in Kansas)
Metal chimney liners

Michigan Chim Flex
39210 West 9 Mile Road
Northville, MI 48176
(800) 289-2446
Metal chimney liners

National Steelcrafters of Oregon
P.O. Box 24910
Eugene, OR 97402
(503) 683-3210
Breckwell (brand) pellet stoves

National Supaflu Systems
P.O. Box 89
Walton, NY 13856
(607) 865-7636
(800) 788-7636
Cast masonry chimney liners

Nighthawk Systems
4835 Centennial Boulevard
Colorado Springs, CO 80919
(800) 880-6788
Nighthawk (brand) CO detectors

Nu-Tec Inc.
Box 908
East Greenwich, RI 02818
(401) 738-2915
Improved woodstoves

Patrick Plastics, Inc.
18 Basaltic Road
Vaughan, Ontario L4K 1G6
Canada
(800) 203-7987
S-Tech (brand) CO detectors

Protech Systems, Inc.
29 Gansevoort Street
Albany, NY 12202
(518) 463-7284
Titanium–stainless steel liners with masonry insulation

Pyro Industries
695 Pease Road
Burlington, WA 98233
(206) 757-9728
Whitfield (brand) pellet stoves

Quantum Group, Inc.
11211 Sorrento Valley Road, Suite A
San Diego, CA 92121
(800) 432-5599
Quantum (brand) CO detectors

Simpson Duravent
P.O. Box 1510
Vacaville, CA 95696
(800) 227-8446
(800) 922-1611 (in California)
Metal chimney liners

Solid/Flue Chimney Systems, Inc.
370 100th Street
Byron Center, MI 49315
(800) 444-3583
Cast masonry chimney liners

Travis Industries
10850 117th Place N.E.
Kirkland, WA 98033
(206) 827-9505

Travis (brand) pellet stoves; improved woodstoves

Vermont Castings, Inc.
16 Airpack Road, Suite 3
West Lebanon, NH 03784
(802) 234-2300
Vermont Castings (brand) and Waterford (brand) pellet stoves; improved woodstoves

Winrich International Group
P.O. Box 51
Bristol, WI 53104
(800) 755-8403
(414) 857-7542 (in Wisconsin)
Winrich (brand) pellet stoves

Z-Flex
20 Commerce Park North
Bedford, NH 03110-6911
(800) 654-5600
(603) 669-5136 (in New Hampshire)
(603) 669-0309 (fax)
Metal chimney liners and venting equipment

Books and Publications

The Fireplace Book, by the editors of *Masonry Construction Magazine.* Available from the Aberdeen Group, 426 South Westgate, Addison, IL 60101; (800) 323-3550.

Heating with Wood (DOE/CS-0158). 1986. U.S. Department of Energy. Available from EPA Public Information Center, 401 M Street S.W., Washington, DC 20460; or from EPA regional offices. Concise, thorough guide to fireplaces, woodstoves, and using wood for fuel.

Indoor Air Quality (Biological Contaminants, Volatile Organic Compounds, Improving Ventilation)

Organizations

Allergy and Asthma Network/Mothers of Asthmatics, Inc.
3554 Chain Bridge Road, Suite 2000
Fairfax, VA 22030
(800) 878-4403

American Academy of Allergy and Immunology
611 East Wells Street
Milwaukee, WI 53202
(800) 822-2762

American Society of Heating, Refrigerating, and Air-Conditioning Engineers
1791 Tullie Circle, N.E.
Atlanta, GA 30329
(800) 527-4723

American Society of Home Inspectors
1735 North Lynn Street, Suite 950
Arlington, VA 22209-2022
(703) 524-2008

The Carpet and Rug Institute
Box 2048
Dalton, GA 30722
(706) 278-3176

Indoor Air Division
Office of Air and Radiation
U.S. Environmental Protection Agency
401 M Street, S.W.
Washington, DC 20460
(793) 308-8470

Formaldehyde Institute
1330 Connecticut Avenue, N.W.
Washington, DC 20036
(202) 659-0060

Hardwood Plywood Manufacturers Association
1825 Michael Faraday Drive
Reston, VA 22090
(703) 435-2900

The Home Ventilating Institute
30 West University Drive
Arlington Heights, IL 60004
(708) 394-0150

National Air Duct Cleaners Association
1518 K Street, N.W., Suite 503
Washington, DC 20005
(202) 737-2926

National Association of Home Builders Research Center
400 Prince George's Boulevard
Upper Marlboro, MD 20772-8731
(301) 249-4000

National Association of Waterproofing Contractors
23625 Commerce Park Road, Suite 206
Beechwood, OH 44122
(800) 245-6292

National Jewish Center for Immunology and Respiratory Medicine
1400 Jackson Street
Denver, CO 80206
(800) 222-5864 (800-222-LUNG)
(303) 388-4461 (in Colorado)

345

Product Manufacturers and Suppliers

Building Materials

Air Vent, Inc.
4801 North Prospect Road
Peoria Heights, IL 61614
(800) 247-8368
(309) 688-5020 (in Illinois)
Continuous ridge vents

Akzo Industrial Systems Company
Suite 318
Ridgefield Business Center
Ridgefield Court
Asheville, NC 28802
(704) 665-5050
Enkadrain (brand) subsurface drainage
matting

B-Dry System
1341 Copley Road
Akron, OH 44320
(800) 321-0985
(800) 237-7413 (in Ohio)
Nationwide franchised basement-water-
proofing contractors, specializing in interior footing drains

Beaver Industries
890 Hersey Street
St. Paul, MN 55114
(800) 828-2947
(612) 644-9933 (in Minnesota)
Downspout underground drainage containers and other basement-drainage
systems

Benjamin Obdyke
John Fitch Industrial Park
Warminster, PA 18974
(215) 672-7200
Continuous ridge vents

Cobra Ventilation Co.
1361 Alps Road
Wayne, NJ 07470
(800) 688-6654
Continuous ridge vents

Browning Metal Products
4480 North Prospect Road
Peoria, IL 61614
(800) 841-8970
Complete line of residential ventilation
products, informative catalog

Cor-A-Vent Inc.
16250 Petro Drive
Mishawaka, IN 46544
(800) 837-8368
(219) 255-1910 (in Indiana)
Continuous ridge vents

Dow Chemical Company
Midland, MI 48674
(800) 258-2436
Thermadry (brand) insulated drainage
panels

Geotech Systems
100 Powers Court
Sterling, VA 22170
(703) 450-2366
Geotech (brand) insulated drainage
panels

Lomanco
P.O. Box 519
Jacksonville, AR 72076
(800) 643-5596
(501) 982-6511 (in Arkansas)
Continuous ridge vents

Medite Corporation
P.O. Box 4040
Medford, OR 97501
(503) 773-2522
Formaldehyde-free medium-density
fiberboard

Monsanto Company
2381 Centerline Industrial Drive
St. Louis, MO 63146
(800) 325-4330
Hydraway (brand) subsurface geocom-
posite drainage systems

O-Well Products
80 Enterprise Road
Hyannis, MA 02601
(800) 356-9935
Downspout underground drainage containers

Paints and Coatings
Aspen Paints
1128 Southwest Spokane Street
Seattle, WA 98134
(206) 682-4603
401 Vapor Guard (brand) latex wall primer

Benjamin Moore
51 Chestnut Ridge Road
Montvale, NJ 07645
(201) 573-9600
260-00 Moore Craft Vapor Barrier (brand) latex wall primer

Glidden Paint Co.
925 Euclid Avenue
Cleveland, OH 44115
(216) 344-8000
Insul-Aid (brand) vapor-proof latex wall primer, Spred 2000 VOC-free interior latex wall paint

Miller Paints
317 Southeast Grand Avenue
Portland, OR 97214
(503) 233-4491
1545 Vapor-lok (brand) latex wall primer

Palmer Industries
10611 Old Annapolis Road
Frederick, MD 21701
(301) 898-7848
86001-Seal Vapor Barrier (brand) nontoxic modified latex primer

Rodda Paint
6932 Southwest Macadam
Portland, OR 97219
(503) 245-0788
7900 Vapor Block (brand) latex wall primer

Sherwin Williams, Inc.
Cleveland, OH 44101
(800) 321-8194
(800) 362-0903 (in Ohio)
Vapor Barrier 154-6407 latex wall primer

United Gilsonite Laboratories
P.O. Box 70
Scranton, PA 18501
(800) 845-5227
Drylok (brand) cementitious waterproofing paint

Climate and Air-Cleaning Equipment
Abbaka
435 23rd Street
San Francisco, CA 94107
(800) 548-3930
Range hoods

Amana Refrigeration Inc.
Amana, IA 52204
(319) 622-5511
Range hoods

American Aldes
4537 Northgate Court
Sarasota, FL 34234
(813) 351-3441
Heat-recovery ventilators

Broan Manufacturing Company
P.O. Box 140
Hartford, WI 53027
(414) 673-4340
Electronic air cleaners; humidifiers; range hoods, ceiling fans, attic fans, and whole-house fans

Carrier Corporation
P.O. Box 4808
Syracuse, NY 13221
(800) 227-7437 (800-CARRIER)
(315) 432-6000 (in New York)
Air-conditioning and humidification systems, sealed-combustion heating units, air cleaners

Conservation Energy Systems
2525 Wentz Avenue
Saskatoon, Saskatchewan S7K 2K9
Canada
(800) 667-3717
Heat-recovery ventilators

Dacor
950 South Raymond Avenue
Pasadena, CA 91109
(818) 799-1000
Range hoods

Dri-Steem Humidifier Co.
14949 Technology Drive
Eden Prairie, MN 55344
(800) 328-4447
(612) 949-2415 (in Minnesota)
Humidifiers

Dunkirk
85 Middle Road
Dunkirk, NY 14048
(716) 366-5500
Sealed-combustion heating units

DuroDyne Corp.
130 Route 110
Farmingdale, NY 11735
(800) 899-3876
Heat-recovery ventilators

Casablanca Fan Co.
450 Baldwin Park Boulevard
City of Industry, CA 91746
(818) 369-6441
Ceiling fans

Emerson Electric
8400 Pershall Road
Hazelwood, MO 63042
(800) 325-1184
(314) 595-2500
Ceiling fans, whole-house fans, humidifiers

Energetechs
P.O. Box 184
Victor, MT 59875

(406) 642-3950
Ventilating system designers, heat-recovery ventilators

Energy Federation
14 Tech Circle
Natick, MA 01760
(800) 876-0660
Ventilating system designers, heat-recovery ventilators

Envirocaire
747 Bowman Avenue
Hagerstown, MD 21740-6871
(800) 332-1110
HEPA residential air cleaners

Environment Air Ltd.
P.O. Box 10
Cocagne, New Brunswick E0A 1K0
Canada
(506) 576-6672
Heat-recovery ventilators

Fantech Inc.
1712 Northgate Boulevard
Sarasota, FL 34234
(800) 747-1762
Low-noise range hoods

Faber
P.O. Box 435
Wayland, MA 01778
(508) 358-5353
Range hoods

Fasco Industries
810 Gillespie Street
Fayetteville, NC 28302
Ceiling fans, attic fans, and whole-house fans

Honeywell Inc.
Honeywell Residential Division
1985 Douglas Drive North
Golden Valley, MN 55422-4386
(800) 345-6770
Home climate systems, heat-recovery ventilators, air cleaners

Hunter Fan Co.
2500 Frisco Avenue
Memphis, TN 38114
(800) 252-2112
Ceiling fans, air cleaners

Jenn-Air
3034 Shadeland
Indianapolis, IN 46226-0901
(317) 545-2271
Range hoods

Lennox Industries
P.O. Box 799900
Dallas, TX 75379
(800) 453-6669
Sealed-combustion heating units, humidifiers

Nutech Energy Systems Inc.
511 McCormick Boulevard
London, Ontario N5W 4C8
Canada
(519) 457-1904
Heat-recovery ventilators

Nutone
Redbank and Madison Roads
Cincinnati, OH 45227
(800) 543-8687
(800) 582-2030 (in Ohio)
Ceiling fans, attic fans, and whole-house fans; range hoods

Pollenex/Rival Company
800 East 101st Terrace
Kansas City, MO 64131
(816) 943-4100
Air cleaners

Pure Air Systems
Plainfield, IN 46168
(800) 869-8025
(317) 859-9135
HEPA central air cleaners

Raydot, Inc.
145 Jackson Avenue
Cokato, MN 55321
(800) 328-3813
Heat-recovery ventilators

Research Products Corp.
P.O. Box 1467
Madison, WI 53701
(800) 334-6011
Heat-recovery ventilators, humidifiers

Shelter Supply
1325 East 79th Street
Minneapolis, MN 55425
(612) 854-4266
Ventilating system designers, heat-recovery ventilators

Stirling Technology, Inc.
P.O. Box 2633
9 Factory Street
Athens, OH 45701
(800) 535-3448
Heat-recovery ventilators

Tamarack Technologies, Inc.
P.O. Box 490
West Wareham, MA 02576
(800) 222-5932
Low-noise bathroom and whole-house exhaust fans, ventilating equipment

Thermador
5119 District Boulevard
Los Angeles, CA 90040
(213) 562-1133
Range hoods

Therma-Stor Products
P.O. Box 8050
Madison, WI 53708
(800) 533-7533
Heat-recovery ventilators

United Technologies/Carrier
7310 West Morris Street
Indianapolis, IN 46231
(317) 481-5723
Heat-recovery ventilators, air cleaners

Venmar Ventilation Inc.
1715 Haggerty
Drummondville, Quebec J2C 5P7
Canada
(819) 477-6226
Heat-recovery ventilators

Vent-A-Hood
1000 North Greenville Avenue
Richardson, TX 75080
(214) 235-5201
Range hoods

Viking Range Corp.
P.O. Drawer 956
Greenwood, MS 38940
(601) 455-1200
Range hoods

Whirlpool Corp.
Benton Harbor, MI 49022
(800) 253-1301
(800) 632-2243 (in Michigan)
(800) 253-1121 (in Alaska and Hawaii)
Range hoods, dryer vents, humidifiers

White/Westinghouse
6000 Perimeter Drive
Dublin, OH 43017

(800) 254-0600
Range hoods, dryer vents, humidifiers,
air cleaners

Monitors and Testers
Air Quality Research
P.O. Box 14063
Research Triangle Park, NC 27709
(919) 941-5509
AQR (brand) indoor formaldehyde monitors

Bacharach, Inc.
625 Alpha Drive
Pittsburgh, PA 15238
(412) 963-2624
GMD Systems (brand) formaldehyde
dosimeter badges

Books and Publications

Directory of State Indoor Air Contacts. Public Health Foundation, 1220 L Street, N.W., Washington, DC 20005; (202) 898-5600. Free booklet listing state agency contacts for individual air quality issues, regional offices of EPA and OSHA, and national hot lines for EPA, CSPC, and the U.S. Department of Health and Human Services.

The Healthy House. John Bower. Carol Communications/Lyle Stuart, 600 Madison Avenue, New York, NY 10022. 1989. Encyclopedic overview of health aspects of building materials and products.

Healthy House Building: A Design and Construction Guide. John Bower. The Healthy House Institute, 7471 North Shiloh Road, Unionville, IN 47468. 1993. Detailed guide to building an actual house using nonpolluting, nonallergenic construction materials and products.

Indoor Allergens: Assessing and Controlling Adverse Health Effects. Andrew M. Pope, Roy Patterson, Harriet Burge, editors. Institute of Medicine, National Academy Press. 1993. Medically oriented overview of indoor allergens, testing methods, and remedies.

The Inside Story: A Guide to Indoor Air Quality (EPA/400/1-88/004). 1988. Available from EPA Public Information Center, 401 M Street, S.W., Washington, DC 20460; (202) 382-4355; or from EPA regional offices. Simple overview of basic residential air quality hazards, remedies, and related issues.

Introduction to Indoor Air Quality: A Reference Manual (EPA/400-3-91/003). 1991. Available from EPA Public Information Center, 401 M Street, S.W., Washington, DC 20460; or from EPA regional offices. Supplement to *A Self-Paced Learning Module* (EPA/400-3-91/002), containing exhaustive support information.

Introduction to Indoor Air Quality: A Self-Paced Learning Module (EPA/400-3-91/002). 1991. Available from EPA Public Information Center, 401 M Street, S.W., Washington, DC 20460; or from EPA regional offices. Detailed learning document for environmental health professionals, focusing on residential indoor air quality issues except radon and asbestos.

The Natural House Book. David Pearson. Simon and Schuster/Fireside, Simon and Schuster Building, Rockefeller Center, 1230 Avenue of the Americas, New York, NY 10020. 1989. Inspirational guide to residential building using ecologically harmonious materials and techniques.

Report to Congress on Indoor Air Quality. Vol. 2: *Assessment and Control of Indoor Air Pollution* (EPA/400/1-89/001C). 1989. Available from EPA Public Information Center, 401 M Street, S.W., Washington, DC 20460; or from EPA regional offices. Overview of U.S. government research on indoor air quality issues.

Residential Air-Cleaning Devices: A Summary of Available Information (EPA 400/1-90-002). 1990. Available by calling EPA Publications Requests at (202) 382-4355; or from EPA regional offices. Concise descriptions and facts on main categories of portable and central residential air cleaners.

Controlling Noise

Organizations

U.S. Department of Housing and Urban Development
HUD User's Document Reproduction Service
P.O. Box 6091
Rockville, MD 20850
(301) 251-5154
(800) 245-2691

Product Manufacturers and Suppliers

Akzo Industrial Systems Company
Suite 318
Ridgefield Business Center
Ridgefield Court
Asheville, NC 28802
(704) 665-5050
Under-carpet matting

Armstrong World Industries
P.O. Box 3001
Lancaster, PA 17604
(717) 397-0611
Soundproof partitions, wall and floor coverings, ceiling systems

CertainTeed
P.O. Box 860
Valley Forge, PA 19482
(800) 835-2522
Sound control insulation and resilient
channels

Homasote Company
Box 7240
West Trenton, NJ 08628-0240
(800) 257-9491
(609) 883-3300 (in New Jersey)
(609) 530-1584 (fax)
Sound-insulating fiberboard and other
noise-control components

Manville-Schuller, Inc.
P.O. Box 5108
Denver, CO 80217
(800) 654-3103
Sound-insulation materials

Owens-Corning Fiberglas Corp.
Fiberglas Tower
Toledo, OH 43659
(800) 438-7465
Sound-insulation materials

Sioux Chief
P.O. Box 397
Peculiar, MO 64078
(800) 821-3944
Water hammer arrestors

Watts Co.
815 Chestnut Street
North Andover, MA 01845-6098
(508) 688-1811
Water hammer arrestors, pressure-reducing valves

Books and Publications

Airborne Sound Transmission Loss Characteristics of Wood-Frame Construction.
F. F. Rudder, Jr. Forest Products Laboratory General Technical Report EPL-43 (U.S.
Department of Agriculture). Madison, WI: Forest Products Laboratory. 1985. Summary of available data on airborne sound transmission loss properties of wood-
frame construction and evaluation of methods for predicting loss.

*A Guide to Airborne, Impact, and Structureborne Noise Control in Multifamily
Dwellings.* HUD Document Order Number 7501046. Available from HUD User's Document Reproduction Service, P.O. Box 6091, Rockville, MD 20850; (800) 245-2691 or
(301) 251-5154.

Gypsum Construction Handbook. 1992. United States Gypsum Company, 101 South
Wacker Drive, Chicago, IL 60606-4385. Contains a chapter on noise-resistant wallboard construction techniques.

Protecting Your Home from Fire

Organizations

*International Association of Fire
Chiefs*
1329 18th Street N.W.
Washington, DC 20036
(202) 833-3420

*National Fire Protection
Association*
1 Battery Park March
Quincy, MA 02269-9101
(617) 770-3000

Product Manufacturers and Suppliers

Ansul, Inc.
1 Stanton Street
Marinette, WI 54143
(800) 862-6785
Sentry (brand) fire extinguishers

BRK Brands, Inc.
3901 Liberty Street Road
Aurora, IL 60504
(800) 323-9005
Fire extinguishers, First Alert (brand)
smoke detectors

Maple Chase Co.
2820 Thatcher Road
Downers Grove, IL 60515
(708) 963-1550
Firex (brand) smoke detectors

General Fire Extinguisher Corp.
1685 Shermer Road
Northbrook, IL 60062
(708) 272-7500
General (brand) fire extinguishers

Halo
400 Busse Road
Elk Grove Village, IL 60007
(800) 323-8705
I.C.-rated recessed ceiling lighting fixtures

Juno
2001 South Mount Prospect Road
Des Plaines, IL 60017
(800) 323-5068
I.C.-rated recessed ceiling lighting fixtures

Walter Kidde
1394 South Third Street
Mebane, NC 27302
(800) 654-9677
Kidde (brand) fire extinguishers, smoke
detectors

Lightolier
100 Lighting Way
Secaucus, NJ 07096
(800) 223-0726
I.C.-rated recessed ceiling lighting fixtures

Scientific Component Systems
2651 Dow Avenue
Tustin, CA 92680
(714) 730-3555
I.C.-rated recessed ceiling lighting fixtures

Books and Publications

Don't Get Burned: A Family Fire-Safety Guide. Gary and Peggy Glenn. Aames-Allen
Publishing Co., Huntington Beach, CA 92648. 1982. Advice and strategies for household fire protection and emergency fire-escape procedures.

*Dr. Frank Field's Get Out Alive: Save Your Family's Life with Fire Survival Tech-
niques.* Frank Field and John Morse. Random House, New York, NY 10022. 1992.
Concise, easy-to-follow directions for avoiding and escaping a household fire.

Upgrading Old Wiring

Organizations

*International Association of
Electrical Inspectors*
901 Waterfall Way, Number 602
Richardson, TX 75080
(800) 786-4234
(214) 235-1455 (in Texas)

*National Electrical Manufacturers'
Association*
210-1 L Street, N.W.
Washington, DC 20037
(202) 457-8400

Product Manufacturers and Suppliers

Square D Company
Executive Plaza
Palatine, IL 60067
(708) 397-2600
Home wiring systems, electrical wiring
components

Leviton Manufacturing Company
59-25 Little Neck Parkway
Little Neck, NY 11362
(718) 229-4040
(800) 832-9538

Home wiring systems, electrical wiring
components

The Wiremold Company
60 Woodlawn Street
West Hartford, CT 06133-2500
(800) 621-0049
(203) 233-6251 (in Connecticut)
Residential surface wiring systems

Books and Publications

Basic Wiring. By the editors of Time-Life Books. Time-Life Books, Alexandria, VA.
1994. Beginner's guide to essential residential wiring repairs and improvements.

Do Your Own Wiring. K. E. Armpriester. Sterling Publishing Company, New York,
NY. 1991. Electricity and residential wiring basics for amateurs.

Electrical Wiring for the Home. Jeff Markell. Prentice-Hall, Englewood Cliffs, NJ
07632. 1991. Thorough guide to residential wiring for professionals.

Electrical Wiring—Residential. Ray C. Mullin. Delmar Publishers, Inc., Albany, NY.
11th ed. 1993. Clearly written, thorough room-by-room guide to residential wiring in
compliance with latest version of National Electrical Code rules. Revised regularly.
For professionals and amateurs.

Illustrated Guide to the National Electrical Code, 1993. John E. Traister. Craftsman
Book Co., Carlsbad, CA 92018. 1993. Wiring methods and other electrical installations updated to comply with 1993 National Electrical Code. Code references included. For amateurs and professionals.

Modern Residential Wiring. Harvey N. Holzman. Goodheart-Willcox Co., South Hol-
land, IL 60473. 1993. Revised edition of thorough vocational text.

Modern Wiring Practice. W. E. Steward and T. A. Stubbs. Butterworth-Heinemann, Newton, MA 02158. 1992. Thorough electrical instruction for vocational students and professionals.

National Electrical Code. Published yearly by the National Fire Protection Association, Quincy, MA. Complete code is revised every three years. For professionals. *Wiring Simplified: Based on 1993 Code.* H. P. Richter and W. C. Schwan. Park Publishing, Inc., Somerset, WI 54025. 1993. Up-to-date edition of popular paperback containing sound electrical instruction for amateurs distilled from the authors' larger book, *Practical Electrical Wiring.* For professionals.

Avoiding Falls and Other Injuries

Organizations

*National Kitchen and Bath
Association*
687 Willow Grove Street
Hackettstown, NJ 07840
(908) 852-0033

Product Manufacturers and Suppliers

American Standard
U.S. Plumbing Products
1 Centennial Plaza
Piscataway, NJ 08854-3996
(800) 223-0068
Temperature-compensating valves

Delta Faucet Co.
55 East 111th Street
Indianapolis, IN 46280
(800) 345-3352 (800-345-DELTA)
Temperature-compensating valves

Eljer Plumbingware
P.O. Box 879001
Dallas, TX 75287-9001
(214) 407-2600
Temperature-compensating valves

Euromix Group
P.O. Box 3188
Blaine, WA 98230
(800) 663-8771
Temperature-compensating valves

Grohe Co.
241 Covington Drive
Bloomingdale, IL 60108
(708) 582-7711
Temperature-compensating valves

Kohler Co.
Kohler, WI 53044
(800) 456-4537 (800-4-KOHLER)
Temperature-compensating valves

Moen
25300 Al Moen Drive
North Olmstead, OH 44070
(216) 962-2000
Temperature-compensating valves

Peerless Products
P.O. Box 2469
Shawnee Mission, KS 66201
(800) 279-9999
Temperature-compensating valves

Price Pfister, Inc.
13500 Paxton Street
Pacoima, CA 91331
(818) 896-1141
Temperature-compensating valves

Scald-Safe (brand) temperature-compensating faucet aerators, tub spouts, and shower arm regulators

Resource Conservation, Inc.
95 Commerce Road
Stamford, CT 06902
(203) 964-0600

Books and Publications

Emergencies in the Home. Arthur R. Couvillon. Information Guides, Hermosa Beach, CA 90254. 1991. Quick response measures for household accidents.

Home Safety & Security. By the editors of Time-Life Books. Alexandria, VA 22314. 1990. Well-illustrated, reasonably thorough do-it-yourself guide to increasing indoor and outdoor household safety and installing security systems.

One Hundred, Twenty-Five Ways to Handle and Prevent Home Hazards and Emergencies. Gary D. Branson. Hearst Books, New York, NY 10019. 1994. Specific responses to the most common household dangers and accidents.

Childproofing

Organizations

American Academy of Orthopaedic Surgeons
222 South Prospect Avenue
Park Ridge, IL 60068
(800) 346-2267
(708) 823-7186 (in Illinois)

American Academy of Pediatrics
141 Northwest Point
P.O. Box 927
Elk Grove Village, IL 60009-0927
(708) 228-5005

Juvenile Products Manufacturers Association, Inc.
2 Greentree Center, Suite 225
P.O. Box 955
Marlton, NJ 08053
(609) 985-2878

National Fire Protection Association
Public Affairs
P.O. Box 9101
1 Battery Park March
Quincy, MA 02269-9101
(617) 770-3000

National Institute of Standards and Technology
B321 Research Building
Gaithersburg, MD 20899-0001
(301) 975-5648

National SafeKids Campaign
111 Michigan Avenue, N.W.
Washington, DC 20010-2970
(202) 884-4993

National Safety Council
1121 Spring Lake Drive
Itasca, IL 60143-3201
(708) 285-1121

U.S. Consumer Product Safety
Commission
Washington, D.C. 20207
(800) 638-2772

Toy Manufacturers Association
P.O. Box 866
Madison Square Station
New York, NY 10159
(212) 675-1141

Window Covering Safety Council
355 Lexington Avenue
New York, NY 10017
(800) 506-4636

Product Manufacturers and Suppliers

Banix Corp.
11835 Carmel Mountain Road
San Diego, CA 92128
(619) 673-1863
Banister guards

Fisher-Price
636 Girard Avenue
East Aurora, NY 14052
(716) 687-3000
Manufacturer of safe products, including toys and furniture, for children

Gerry Baby Products
1500 East 128th Street
Denver, CO 80241
(800) 626-2996
Manufacturer of safe products, including carriers, for children and infants

KinderGard Corp.
14822 Venture Drive
Dallas, TX 75234
(214) 243-7101

Electrical safety products, childproof latches

Perfectly Safe
7245 Whipple Avenue, N.W.
North Canton, OH 44720
(216) 494-2323
Mail-order source of childproofing and household accident-prevention products

Safety 1st, Inc.
210 Boyleston Street
Chestnut Hill, MA 02167
(800) 962-7233
Manufacturer of children's products (largest distribution in U.S.)

The Safety Zone
Hanover, PA 17333
(800) 999-3030
Mail-order source of childproofing and household accident-prevention products

Books and Publications

Children's Play Yards. By the editors of Sunset Books and *Sunset Magazine.* Lane Publishing Company, Menlo Park, CA 94205. 1989. Outdoor building projects and advice on creating a child-safe backyard environment.

The Childwise Catalog. Jack Gillis, Mary Ellen R. Fise, and Consumer Federation of America. Harper and Row, Publishers, New York, NY 10022. 1990. Compendium of general advice on children's products, safety, accident prevention, health, and child care issues.

Keeping Your Children Safe: A Practical Guide for Parents. Bettie B. Youngs. Westminster/John Knox, Louisville, KY 40202. 1992. Thorough advice on child safety and accident prevention.

The Kids' Collection. Jane Smolik. The Globe Pequot Press, Chester, CT 06412. 1990. Directory of catalogs from more than five hundred retailers specializing in children's products, including health and safety items.

Making Your Home Child-Safe. Don Vandervort and the editors of Sunset Books. Lane Publishing Company, Menlo Park, CA 94025. 1988. Room-by-room improvements to increase child safety in a home and play yard.

Safe Kids: A Complete Child Safety Handbook and Resource Guide for Parents. Vivian K. Fancher. John Wiley and Sons, Inc., New York, NY 10158. 1991. Thorough coverage of household safety and child protection.

Creating a Barrier-Free Home

Organizations

Able Data
8455 Colesville Road, Suite 935
Silver Spring, MD 20910
(800) 227-0216
Consultants for the National Institute on Disability and Rehabilitation Research, U.S. Department of Education

American Association of Retired Persons
601 E Street, N.W.
Washington, DC 20049
(202) 434-2277

Association of Home Appliance Manufacturers
20 North Wacker Drive
Chicago, IL 60606
(312) 984-5800

Product Manufacturers and Suppliers

Amana Refrigeration, Inc.
Amana, IA 52204
(800) 843-0304
Cabinet-depth refrigerators, shallow-cavity microwave ovens, cooktops

American Standard
Piscataway, NJ 08854-3996
(800) 223-0068
Wheelchair-accessible lavatories, toilets

Bathease
2537 Frisco Drive
Clearwater, FL 34621
(813) 791-6656
Barrier-free bathtubs

The Braun Corporation
1014 South Monticello
P.O. Box 310
Winamac, IN 46966
(800) 843-5438
(219) 946-6153 (in Indiana)
Roll-in showers

Eljer
17120 Dallas Parkway
Evergreen Center
Dallas, TX 75248
(800) 423-5537
Accessible bathroom systems

Frigidaire Company
6000 Perimeter Drive
Dublin, OH 43017
(614) 792-4733
Wheelchair-accessible kitchen ranges,
appliances

Honeywell
Residential Building Controls
1985 Douglas Drive North
Golden Valley, MN 55422-3992
(800) 345-6770
Enlarged-scale, raised-print thermostats

Lindustries, Inc.
P.O. Box 295
Auburndale, MA 02166
(617) 237-8177
Manufacturers of Leveron (brand) lever
doorknobs

Maytag Co.
1 Dependability Square
Newton, IN 46280
(317) 848-7933

Safetek International, Inc.
P.O. Box 23
Melbourne, FL 32902
(407) 952-1300
Support bars and handrails

Sterling Plumbing Group, Inc.
1375 Remington Road
Shaumburg, IL 60173-4898
(708) 843-5400
Barrier-free showers, tubs, and bathroom fixtures

Whirlpool Corporation
2000 North M-63
Benton Harbor, MI 49022-0692
(800) 253-1301
Accessible bathroom systems

Books and Publications

Building for a Lifetime. Margaret Wylde, Adrian Barron-Robbins, and Sam Clark. Taunton Press, Inc., 63 South Main Street, Box 5506, Newtown, CT 06470-5506. 1994. Complete, detailed guide to building accessible living environments. Thorough bibliography and lists of resources.

The Complete Guide to Barrier-Free Housing. Gary D. Branson. Betterway Publications, Inc., P.O. Box 219, Crozet, VA 22932; (804) 823-5661. 1991. Thorough overview of residential remodeling techniques and products that enhance living for the elderly and physically handicapped. Exhaustive directory of information sources, organizations, product manufacturers, and mail-order retailers.

Directory of Accessible Building Products. National Home Building Association Research Center, 400 Prince George's Boulevard, Upper Marlboro, MD 20772-8731; (301) 249-4000. Guide to available products for homeowners with special accessibility needs.

INDEX

acetone (methyl ethyl ketone), 148, 153
actinolite, 3–4
adhesives, 24–25
  biological contaminants and, 126, 136
  in encapsulating asbestos, 19
  noise and, 190–91, 193, 195
  radon and, 63
adjuster knobs, 94
adults:
  combustion products and, 69–70, 72, 95
  lead and, 28–29, 34
  radon and, 59
  aeration, 65–67
Africa, asbestos deposits in, 4–5
Agency for Toxic Substances and Disease Registry (ATSDR), 32
air chambers, 201–2
air cleaners:
  airflow rates of, 147
  biological contaminants and, 142–47
  CADRs of, 146–47
  combustion products and, 78–79, 142
  console, 143–44
  electret, 143–45
  electronic, 143–47
  electrostatic, 143–44, 147
  electrostatic precipitating, 143–44
  finding information on, 347–50
  illustrations of, 144–46
  mechanical, 143, 147
  negative-ionizing, 143, 146

ratings of, 146–47
selection of, 146–47
air conditioners, 240
  biological contaminants and, 113, 121, 137
  cleaning coils of, 80
  combustion products and, 78, 80
  finding information on, 347–50
  noise and, 188, 207–8
  radon and, 57–58
  ventilation and, 175
Air-Conditioning and Refrigeration Institute Standards, 141
air dampers, 92
airflow regulators, 142
air intakes, 104
air sampling, 10
air shafts, 170
air terminals, 230–33
alarm systems, 213
alkyd-based enamels, 47
allergic rhinitis (hay fever), 111
allergies, xv, 329
  biological contaminants and, 111–12, 114, 116–18, 120–21, 140, 143
  combustion products and, 71
  humidifiers and, 140
  MCS and, 329
alpha particles, 51–52, 58, 66–67
alpha-track detectors, 58–60
alternating current (AC), 73–74, 236–38, 327
alumina, 49

361

floor-leveling compound, 20
floor polishers, 25
floors:
asbestos in, 6, 9, 12, 20–24
and avoiding injuries, 285, 287, 295, 298
barrier-free homes and, 316–18, 320
biological contaminants and, 116–18, 120, 122, 124–27, 132, 134–35
childproofing and, 302, 304–5
combustion products and, 108
fire and, 217, 222
flooring over, 20, 24–25
framed, 195–97
and inspecting household wiring, 250
lead removal and, 41–42, 46
noise and, 184–86, 188, 190, 193–95, 197, 200, 204, 208
radon and, 50, 62–64
removal of, 20–25
silencing pipes running through, 204
slip-resistant, 298
underlayment of, 25
vapor barriers over, 126
ventilation and, 165, 170, 172
VOCs and, 150–51, 154, 159
flues:
combustion products and, 72, 82, 85–86, 88–90, 92, 103, 105, 108
condensing furnaces and, 103
connecting different types of combustion appliances to, 88, 92
of zero-clearance fireplaces, 105
fluorescent lighting:
in barrier-free homes, 318
repairing of, 272, 275
SAD and, 328–29
foam drainage panels, 135–36
foliage, 130–31, 233
food cans, 30
footing drains, 135–36
forced-air heating systems:
adding air cleaners to, 143, 146–47
adding humidifiers to, 137, 140–41
biological contaminants and, 111, 140–41, 143, 146–47
ventilation and, 179

formaldehyde, 148–52, 155–59
concentrations, standards, and effects of, 155–56
disposal of, 158–59
household sources of, 150–52
testing for, 157–58
uses of, 149–50
ventilation and, 161
foundations:
biological contaminants and, 130–31, 134–37
combustion products and, 82
and inspecting household wiring, 251, 253
noise and, 199, 203–4
radon and, 60, 62, 64
slope of ground near, 130–31
4-phenylcyclohexene, 148, 153
freezers, 321–22
asbestos in, 9
and repairing electrical systems, 277
frost, 138–39
fuel gauges, 94
fuel supply lines, 76–77
fuel tanks, 76, 94
full-spectrum lighting, 328–29
fungi, 111, 115, 147
furnace rooms, 223
furnaces, 139
asbestos in, 8–9, 12
backdrafting and, 162–64
combustion products and, 68, 72–74, 76–80, 84, 88, 92, 95, 100, 102–5, 108
condensing, 103–4
filters of, 144–45
gas, *see* gas furnaces
inspection and maintenance of, 76–80, 88, 92, 100, 102–3
life expectancies of, 103
noise and, 207
oil, *see* oil furnaces
properly sized, 102–3
and repairing electrical systems, 277
sealed-combustion, 103–5
standards for fuel efficiency in, 102
ventilation and, 162–64, 179–80
furniture:
and avoiding injuries, 283–85
in barrier-free homes, 319–20, 325

372

Index

insulation (*cont'd*)
fire and, 214, 216–18
and inspecting household wiring, 254–57
and insuring safety of electrical systems, 241–42, 245
noise and, 188–91, 193–95, 197–200, 203–5, 208
radon and, 64
and repairing electrical systems, 267–68, 273, 275
ventilation and, 161–62, 167, 172–73, 177, 181
VOCs and, 149–51
intermittent (spot) ventilators, 162
International Agency for Research on Cancer, 150
intestines, 27
ionization smoke detectors, 220–21
iron, 40
ironing board covers, 10
isopropanol (rubbing alcohol), 153

joint compound, 12
joists:
and avoiding injuries, 287–88
mounting ducts parallel to, 207
mounting ducts perpendicular to, 208
noise and, 195–97, 203, 205–8
and repairing electrical systems, 261, 274–75
securing pipes parallel to, 203
securing pipes perpendicular to, 203
ventilation and, 177, 180–81
junction boxes:
and insuring safety of electrical systems, 244
and repairing electrical systems, 261, 274
Juvenile Products Manufacturers Association, 307

kerosene:
grades of, 219
VOCs and, 154
kerosene heaters:
backdrafting and, 163

combustion products and, 68, 72–74, 94–95
fire and, 218–19
illustration of, 94
inspection and maintenance of, 94–95
VOCs and, 151
kidneys, 27–28
kindling, 84
kitchen range hoods:
in barrier-free homes, 321
biological contaminants and, 113, 125
down-draft, 167–68, 170–71
illustrations of, 168–71
island-installed, 168–69
noise ratings for, 169–70
peninsula-installed, 168–69
radon and, 57
updraft, 167–68
ventilation and, 167–71
kitchen ranges:
and avoiding injuries, 284
in barrier-free homes, 320–21
biological contaminants and, 113
childproofing of, 304
vents for, 163
see also electric kitchen ranges; gas kitchen ranges
kitchens:
avoiding injuries in, 283–84, 298–300
in barrier-free homes, xv, 311, 318, 320–22
biological contaminants and, 112–17
childproofing of, 302–4, 308
in circuit maps, 278
combustion products and, 68, 72–74, 95–100
fire and, 213, 219–21, 223, 225–26, 228
lead in, 32, 47
radon and, 57
and repairing electrical systems, 277
ventilation and, 161–62, 167–70
VOCs and, 159
knickknack collections, 114, 117
knives:
biological contaminants and, 127, 133
childproofing and, 303
in floor removal, 21–23, 25
in lead removal, 44
Kraft-paper, 122

376

ABOUT THE AUTHOR

FROM 1987 TO 1994, JOHN WARDE WROTE THE weekly columns "Home Improvement" and "Home Clinic" for *The New York Times*. Both columns appeared in other newspapers nationwide. Mr. Warde is also the author of *The New York Times Season-by-Season Guide to Home Maintenance*, and has made numerous appearances on radio and television, including ABC's *Good Morning America*, *Home Show*, and CBS's *This Morning*. A nationally recognized journalist, magazine writer, book and multimedia editor, Mr. Warde is an expert at understanding and solving the home repair problems of everyday homeowners. He lives with his family near Seattle.